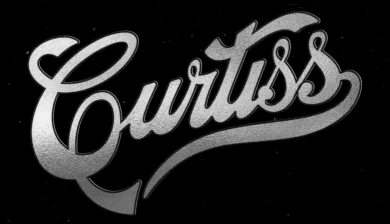

MOTORCYCLES

1902 – 1912

Richard Leisenring Jr.

SCHIFFER
PUBLISHING

4880 Lower Valley Road • Atglen, PA 19310

Cover and title page image: Glenn H. Curtiss on his Hercules V-Twin racer, which set a record at Ormond Beach in January of 1904.

Cover design by Chris Bower
Type set in Capitol Capitals/Scotch Text
ISBN: 978-0-7643-6808-0
Printed in India

Published by Schiffer Publishing, Ltd.
4880 Lower Valley Road
Atglen, PA 19310
Phone: (610) 593-1777; Fax: (610) 593-2002
Email: info@schifferbooks.com
Web: www.schifferbooks.com

For our complete selection of fine books on this and related subjects, please visit our website at www.schifferbooks.com. You may also write for a free catalog.

Schiffer Publishing's titles are available at special discounts for bulk purchases for sales promotions or premiums. Special editions, including personalized covers, corporate imprints, and excerpts, can be created in large quantities for special needs. For more information, contact the publisher.

CONTENTS

ACKNOWLEDGMENTS

No one can actually write a book completely on their own. After helping well over forty authors myself as a researcher for the past forty-five years, I fully understand the importance of that support and the contributions others have to offer to make it complete. With that said, first I need to thank my sainted wife, Jo Anne, for putting up with all of my moaning and groaning, frustrations, and elations over new discoveries these past six years. Her patience in listening to me ramble on about Curtiss and motorcycles was astounding. Not to mention all of the proofreading she cheerfully and involuntarily found herself doing.

Second, several Curtiss experts in their own right have been great in encouraging me and pointing out resources when needed. I am grateful to these gentlemen—Mr. Tom Heitzman, Mr. Dale Axelrod, and Mr. Dale Stoner. I hope you are proud of what has been accomplished.

I would also like to thank these individuals and institutions for their gracious assistance: Joshua Leisenring, Asheville, North Carolina, for his renditions of the logos and emblems of the various Hammondsport-built brands; Hanah Stevens-Burns, Tupper Lake, New York, assistant extraordinaire; James Quint, director, G. H. Curtiss Museum of Local History, Hammondsport, New York; Dick and Sue Rogers, Hackettstown, N.J. for their encouragement and support; and most importantly, my extraordinarily patient editor, Karla Rosenbusch, and the creative staff at Schiffer Publishing who made this project.

Digital archives found on the internet include Newspapers.com; Detroit Public Library, Automotive History Collection, Detroit, Michigan; Google Books; jstor.org; Smithsonian Libraries, library.si.edu; Internet Archive; and openlibrary.edu.

INTRODUCTION

The year 1900 was not only the beginning of a new century, but also a new phase in the history of Hammondsport, Steuben County, New York, home of a local boy with big ambitions—Glenn Hammond Curtiss. This sleepy little village, dependent for years on agriculture and the wine industry, would soon find itself the center of attention in a new and modern industrial era.

In the recent past, scholars had to rely partly on fading memories and questionable stories passed down over the years as fact when trying to document this portion of local history. However, thanks to the electronic age of the twenty-first century, a large amount of actual documentation never seen before is now currently available to the public. Through scanning and digitizing of newspapers, periodicals, legal documents, letters, photographs, and various other materials, a wealth of knowledge is now being brought forth in unprecedented amounts. As a result, it is clear that much of Glenn H. Curtiss's early motorcycle and dirigible years had gone either unnoticed, ignored, or as in some cases, recorded incorrectly. It has taken several years to put a more accurate and completely unbiased history together that will not only dispel a few of the urban legends widely circulated but also give a better, more inclusive picture of Glenn H. Curtiss and his early machines. Also added is a picture of his life leading up to his entrepreneurial ventures to give a better understanding of his background. We also

acknowledge the fact that as time moves forward, other details will come forward to correct the current storyline or add information not currently included.

This is the story of the first major step in Glenn's fascination with speed and mechanical innovations.

It will take you through the birth, life, and demise of one phase of a local industry that would forever have a major effect on the future of society and modern transportation throughout the world.

Sheather Street, Hammondsport, N. Y.

THE EARLY DAYS

Born on May 21, 1878, Glenn Hammond Curtiss, son of Frank—a harness maker by trade—and his wife, Lua (Andrews) Curtiss—an aspiring artist and musician—derived his name from his birthplace, the village and glen of Hammondsport, which is located on the southern tip of Keuka Lake in the Finger Lakes Region of New York State. Shortly after his birth, the family purchased Castle Hill, later referred to as the Curtiss Homestead, a large home and 8 acres of vineyards west of the village, overlooking Hammondsport and Keuka Lake. The following year, Frank's parents, Claudius, a retired Methodist-Episcopal minister, and his wife, Ruth (Bramble) Curtiss, would move in to help with the financial struggles the young family was facing. Glenn's sister, Rutha, would be born here three years later, on February 15, 1881.

With both his father and grandfather passing away within five months of each other shortly before his fifth birthday in 1884, Glenn was now the only male in the household. Since times were hard financially for the family, without a steady income, the family began renting out rooms and doing domestic work. Glenn would start working at the age of ten to help bring in additional income for the family. The many jobs he did when not in school included herding cows, tending vines, and picking grapes in the vineyards behind the Curtiss homestead and surrounding community, pumping the church organ on Sundays, and delivering groceries.

During his childhood, Glenn was an avid outdoorsman and loved hunting and camping. As a sportsman, he enjoyed swimming and ice-skating. However, what he excelled at the most was tinkering and inventing. His inquisitive mind never allowed him to stop figuring out how anything mechanical worked, and he was always looking for ways to improve upon them. It was common to find small mechanical items such as old clocks in various states of repair around the Curtiss

home. To make up for the lack of a male role model in his life, his grandmother Curtiss became the major influence in his upbringing, encouraging his inquisitive nature and helping him develop a strong moral fiber and a rigid work ethic.

Due to Rutha's gradual loss of hearing during an illness at the age of two, Glenn became extremely protective and supportive of her, spending a lot of his free time helping her learn finger spelling and sign language. She would suffer a total loss of her hearing by the age of six. In the early 1890s, when Rutha reached the age to enroll in the Western New York Institution for Deaf Mutes (Rochester School for the Deaf), she, Glenn, and their mother moved temporarily to Rochester, New York, north of Hammondsport, on Lake Ontario. At this time, while the family was not destitute, Lua realized she would need to find steady work and earned a teaching license and a certificate in freehand drawing at the Rochester Mechanics Institute as a means of generating income for the family.[1]

While staying in the poorer part of Rochester to lower expenses, Lua set up a school in an abandoned store next to their apartment to teach the underprivileged children in the area to read and write. She also held sewing classes for the local girls. When possible, Lua taught painting and music to more-affluent residents of the city for extra income. To help with the support of his immediate family and his grandmother back in Hammondsport, Glenn finished his schooling in Rochester after graduating from the eighth grade at the age of fourteen, in order to work steadily at a variety of jobs. This also allowed him to divide his time more evenly between Rochester and Hammondsport. Using some of his earnings to travel by train, Glenn would periodically return to Hammondsport to help maintain the Curtiss homestead.[2]

While living in the city, Glenn held a variety of jobs such as delivering groceries with a makeshift wagon

The Curtiss homestead, known as Castle Hill, ca. 1900

View of the Curtiss home (right forefront), with a view of the village of Hammondsport, ca. 1880

Glenn and Rutha Curtiss, taken in Rochester, New York in 1892

and adult-sized tricycle he built. His sister, Rutha, called it his "Express Wagon," and he also used it to cart her around while on errands. He later purchased a used two-wheel bicycle from a junk dealer and repaired it. For a short time, Glenn also worked for the Eastman Dry Plate Co. (Kodak) as a factory worker. It is believed but not substantiated that he invented a stenciling machine to help productivity in packaging film. During his time there, he developed a love for photography and created his first camera out of an old cigar box. His most notable job, however, would be for the Western Union Telegraph Company. It was here that Glenn found a competitive sport he could excel in—cycle racing. While working as a bicycle messenger delivering telegrams around the city, Glenn and the other delivery boys would hold impromptu races on their bicycles in between assignments. According to his sister, Rutha, Glenn was hard to beat.[3]

Bicycle racing as a sport had become an overnight sensation in the nation when the "bicycle craze" got its start in the late 1880s. This was a result of the introduction of the "safety bicycle," a cycle with two wheels of the same size and utilizing pneumatic tires.

One of the many sports Glenn enjoyed was ice skating. During the winter of 1894, while skating on the Erie Canal in Rochester with a group of other boys, one of the young men fell through the ice. Glenn immediately went to his rescue, pulling him out. Later, when a police officer taking note of the incident asked Glenn why he did it, his only response was "I was the nearest one to him."[4]

When Glenn's mother remarried in April 1895, she moved to Rock Stream, New York, located on Seneca Lake, roughly 25 miles east of Hammondsport. Glenn relocated with her, leaving Rutha in the care of the school. During this time, he would ride his bike between the two villages on weekends to help his grandmother. While on one of these weekend trips, Glenn met C.

Leonard "Tank" Waters, who would become a close friend and partner in several ventures later in life.

In 1896, Glenn found it necessary to move back to the family homestead in Hammondsport to take care of his aging grandmother. In his spare time, Glenn quickly became a local champion bicycle racer on a team known as "the Hammondsport Boys," captained by Tank Waters and sponsored by the village druggist and bicycle dealer, James Smellie.

Glenn won his first major race in 1897 on a borrowed Stearns Racing cycle, and as a result, Smellie used the feat to get the E. C. Stearns & Company of Syracuse, New York, to sell one to Glenn at a discount, with the idea of using Glenn's racing prowess to promote sales. In his correspondence with Stearns promoting the idea, Smellie described Glenn as a "poor zany fellow" who needed the price to be as low as possible.[5]

C. Leonard "Tank" Waters on his bicycle, supported by Glenn H. Curtiss, ca. 1896

Glenn H. Curtiss on his bicycle, supported by an unknown friend, ca. 1896

Promotional photograph of Glenn H. Curtiss on his 1897 Stearns Racing cycle

The Hammondsport Football Team, ca. 1897. This was taken by Glenn (far left); note the camera trigger bulb in his hand. Future Curtiss photographer Harry Benner is in the center.

When not racing, Glenn was involved with the local football team. While working for Smellie repairing bicycles part time, Glenn would put his riding skills and love of photography to work as a traveling photographer for a local studio run by H. E. Saylor. In addition, he worked as a handyman around the village and installed and maintained acetylene gas generators of his own design for lighting in various retail businesses on the square, before electricity was introduced to the area. His bicycle shop would be illuminated by no fewer than ten acetylene-fueled lamps of his own construction.[6]

Glenn and Lena Curtiss on their wedding day, 1898. Glenn took this photo; the camera trigger bulb is hidden in his left hand

On March 7, 1898, just before his twentieth birthday, Glenn married a local girl, Lena Pearl Neff, formerly of Prattsburgh, New York. The two had met four years earlier, while both attended the Methodist-Episcopal Church in Hammondsport during one of his many visits from Rochester and began dating immediately after meeting.[7] Their marriage was considered a bit controversial in the day. While Glenn and Lena were of legal age, the two had been married in secret under another denomination by Reverend C. L. Luther at the Presbyterian Parsonage before telling their parents. The announcement would appear a week later in the *Hammondsport Herald*.

With a growing family to support, Glenn decided to go into business on his own. The financial backing came from his grandmother remortgaging the homestead.[8] Naturally, with bicycles being his major focus at the time, Glenn decided that was where he could make a living, and he opened his first bicycle shop in March 1899 on the village square in Hammondsport.[9] Glenn reportedly had dreams of creating his own brand of bicycle. However, to ensure there was an income when cycle sales were slow, Glenn also sold sewing machines, horse harnesses, and, for a short time, rabbits. He opened a second retail shop 10 miles down the road in the village of Bath the following year.[10] It was at this time Glenn grew his trademark mustache, since he felt it made him look mature and more businesslike.

Glenn's success with his bicycle sales and repairs soon became large enough to drive out the competition in the village, with the jeweler George Lyon discontinuing sales and James Smellie turning over his agencies to Glenn by 1901.

The Curtiss Bicycle Shop in Hammondsport, New York. Glenn Curtiss stands to the far left, ca. 1900.

An advertising photo of Glenn on a bicycle; note his new
mustache (ca. 1900).

James Smellie's Drug & Dry Goods Store, with James Smellie in the center, ca. 1900

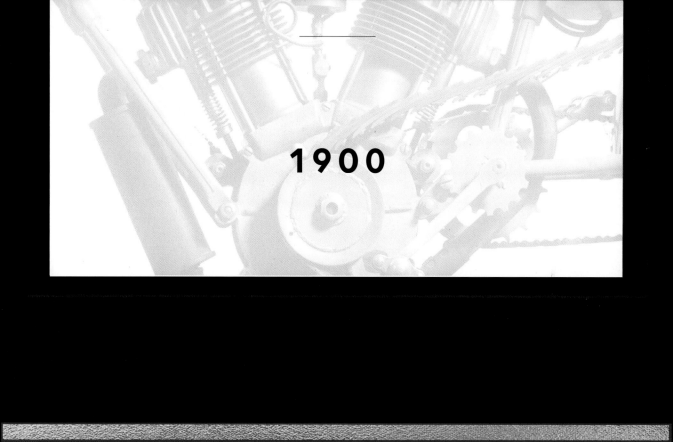

1900

The turn of the twentieth century was truly an age of innovation, and young Glenn would be swept up in it like no other. Three new modes of transportation were simultaneously gaining attention that would dramatically change his life—that of motorcycles, airships, and automobiles. Always looking ahead with his eye on efficiency and speed, Glenn began experimenting on his own motorcycle design. No one knows exactly when Glenn decided to make a motorcycle, but in the summer of 1900, with the help of his wife's uncle, Frank Neff, an accomplished machinist and successful inventor, the process of experimentation began taking place at the Neff machine shop and in the backroom of Glenn's Hammondsport bicycle shop. In mid-October, the pair publicly announced that their one-cylinder machine was about ready for road testing.[1] Using stock engine castings purchased possibly through mail order, their first attempt proved too large, heavy, and overpowered. It was described as having a frame not unlike a bicycle, only heavier, and the engine had a speed of 2,500 revolutions per minute.[2]

Early on, this first engine was thought to be a mail order purchase from the E. R. Thomas Motor Company in Buffalo, New York. It should be pointed out here that the Thomas engine of this power offered in 1900 was one cylinder, 3 horsepower (hp), and 18.5 inches tall and weighed 56 pounds. However, the company had introduced selling castings of the 1.5hp and 3hp engines only in November 1900,[3] their public offering being too late to fit the timeline. At the time of this writing, it is unknown as to where or how exactly these first castings were obtained.

Glenn H. Curtiss or Frank Neff on their first motorcycle, ca. 1900

During the first week in November, Glenn ventured away from home to New York City by train to attend the first major national automobile show at Madison Square Garden.[4] It's not known how long he stayed at the weeklong event, but it was becoming apparent that his dream now extended far beyond just owning a couple of bicycle shops.

Advertising card for Thomas engines and motor bicycles

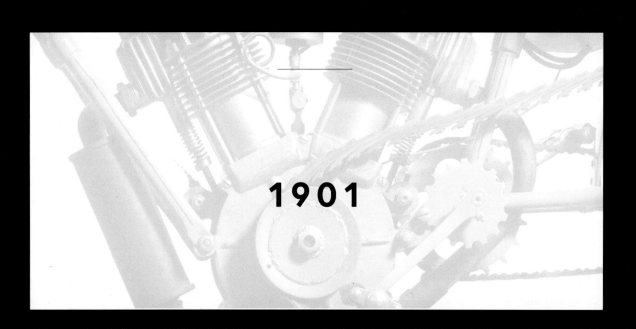

CHAPTER 3

1901

Schiffer Publishing

Ready to Write a Book?

Our authors are as passionate as we are about providing new and intriguing perspectives on a variety of topics, both niche and general. If you have a fresh idea, we would love to hear from you, as we are continually seeking new authors and their work. Visit our website to view our complete list of titles and our current catalogs. Please visit our Author Resource Center on our website for submission guidelines, and contact us at proposals@schifferbooks.com or write to the address below, to the attention of Acquisitions.

Schiffer Publishing Ltd.

A family-owned, independent publisher since 1974, Schiffer has published thousands of titles on the diverse subjects that fuel our readers' passions. Explore our list of more than 5,000 titles in the following categories:

ART, DESIGN & ANTIQUES

Fine Art | Fashion | Architecture | Interior Design | Landscape | Decorative Arts | Pop Culture | Collectibles | Art History | Graffiti & Street Art | Photography | Pinup | Sculpture | Body Art & Tattoo | Antique Clocks | Watches | Graphic Design | Contemporary Craft | Illustration | Folk Art | Jewelry | Fabric Reference

MILITARY

Aviation | Naval | Ground Forces | Figures | Pinup | Transportation | World War I & II | American Civil War | Militaria | Modeling & Collectible | Uniforms & Clothing | Biographies & Memoirs | Unit Histories | Emblems & Patches | Weapons & Artillery

CRAFT

Arts & Crafts | Fiber Arts & Wearables | Woodworking | Quilts | Gourding | Craft Techniques | Leathercraft | Carving | Boat Building | Knife Making | Printmaking | Weaving | How-to Projects | Tools | Calligraphy

TRADE

Lifestyle | Natural Sciences | History | Children's | Regional | Cookbooks | Entertaining | Guide Books | Wildlife | Tourism | Pets | Puzzles & Games | Movies | Business & Legal | Paranormal | UFOs | Cryptozoology | Vampires | Ghosts

MIND BODY SPIRIT

Divination | Meditation | Astrology | Numerology & Palmistry | Psychic Skills | Channeled Material | Metaphysics | Spirituality | Health & Lifestyle | Tarot & Oracles | Crystals | Wicca | Paganism | Self Improvement

MARITIME

Professional Maritime Instruction | Seamanship | Navigation | First Aid/Emergency | Maritime History | The Chesapeake | Antiques & Collectibles | Children's | Crafts | Natural Sciences | Hunting & Fishing | Cooking | Shipping | Sailing | Travel | Navigation

SCHIFFER PUBLISHING, LTD.
4880 Lower Valley Road | Atglen, PA 19310
Phone: 610-593-1777
E-mail: Info@schifferbooks.com
Printed in China

www.schifferbooks.com

Glenn's experimentation continued with motorcycles into 1901. A second machine, propelled by a smaller engine obtained from E. R. Thomas Motor Company, proved to be underpowered. It should be noted that over the next fifty years, the order in which the engines he used has been inadvertently switched. Simply placing the contemporaneous news articles of his experiments in chronological order corrects this.

While frustrated over the outcome of his earlier experiments, Glenn realized that these failures put him one step closer to success, and he began to design his own engine. Since Frank Neff was in the process of setting up his own manufacturing business, Glenn sought out the help of his mechanically talented bicycle-racing friend Charles Kirkham. Charles, having recently completed an education as a mechanical engineer, was working for his father, John, at the Kirkham machine shop and foundry in Taggerts Mills, halfway between Hammondsport and Bath. The result of this cooperation was a lightweight, one-cylinder, 2.5hp engine utilizing ball bearings in an aluminum casing. The finished engine was said to attain 40 miles per hour (mph). Installing this on the second machine in place of the underpowered Thomas engine proved to be the success he was looking for.[1] Throughout his life, Glenn would make it a habit of seeking out others to help him in his ventures, realizing early on that solving problems on his own was not practical.

There is an often-repeated story about Glenn, during all of this trial and error, creating a carburetor out of a tomato soup can and some gauze to get an early engine running properly. While very plausible, the story has never been verified or given a specific date to place it chronologically in his early years.

An early Thomas engine, similar to that used on the second cycle developed by Curtiss. *Photo: Richard Leisenring Jr.*

Charles Kirkham, ca. 1901

The first successful Curtiss Hercules one-cylinder engine unassembled

Glenn's experimentation in 1901 was a bit slower than his enthusiastic and inquisitive nature demanded. His first son, Carlton, was born in March. He suffered from a congenital heart defect, making it difficult for Glenn to put his full attention into his project or his business. Understanding the severity of the illness, Glenn spent as much time as possible with his son. Sadly, Carlton would pass away eleven months later.

It was clear to many in the Hammondsport area that Glenn was onto something big with his business and experiments. The support the locals offered in various ways helped take him to the next step. The community was so interested in Glenn's endeavor that the local newspaper, the *Hammondsport Herald*, would report on almost everything he did over the next few years—from business dealings and plant expansions to racing and speed records. Glenn had no idea that his growing celebrity status was about to put the little village of Hammondsport in the headlines, on the map, and finally in the history books.

As early as 1909, colorful stories would arise about this early life and experimentation, in an attempt by well-meaning friends and associates to fill in the back-story of Glenn's growing fame. This was an area he greatly shied away from, making it difficult to accurately report. According to his sister, Rutha, his modesty was one of his "most interesting characteristics."[2] It is important that several of these examples be presented here, not only to dispel some of the "urban legends" that have grown up around Glenn, but to help understand how much the community revered him as a local hero, to the point of—subconsciously or not—embellishing or inventing facts.

One misconception was that Glenn would ride his bicycle from Rochester to Hammondsport on weekends to help his grandmother. The road distance in the 1890s

Lena and Carlton Curtiss, taken in 1901

was well over 100 miles one way—an exhausting trip considering the poor road conditions and hilly terrain that existed at that time. In actuality, Glenn traveled by train on those visits.[3] It seems that over the years, his trips from Rock Stream by bicycle and those from Rochester were blurred into one. Another tale was how Glenn had reportedly gone to Buffalo, New York, to visit the Pan-American Exposition in 1901, finding his inspiration to build a motorcycle there. We now know that he started a year earlier and, due to his son's illness, never ventured far from home.

Other examples can be found in various early publications, such as that written by Augustus Post, a friend and fellow aviation pioneer. In *The Curtiss Aviation Book* by Glenn H. Curtiss and Augustus Post, published in 1912, Post had James Smellie, a local druggist and early employer of Glenn's, giving him the inspiration to build a motorcycle by suggesting something be created after an exhausting trek up the hill to visit the Curtiss bicycle shop. In reality, there were no hills involved; Smellie's shop was only 100 yards away from Curtiss's, across the level village square. Did Smellie make the initial suggestion? No one knows for sure. To Post's credit, he did write a caveat: "It is hoped that there will be in these introductory chapters—for whose writing be it understood, *Mr. Curtiss is not responsible—.*"

WADSWORTH HOTEL AND R. R. STATION, HAMMONDSPORT, N. Y.

Color postcard of Hammondsport, with vineyards in the background

Two other local stories had Glenn running his first motorcycle into Keuka Lake or a tree during a test run, because he had forgotten to install brakes. A conflicting version had him racing the cycle up the dirt roads on the east side of the lake until it ran out of gas, in order to stop. None of these incidents have ever been verified and may have been a misinterpreted version of an incident that took place in April 1904. As reported in the *Hammondsport Herald*, a potential customer, Leonard Whittier, took a Curtiss motorcycle for a test ride and, not knowing how to operate the cycle, would hit a telephone pole while trying to stop.[4]

The first biography on Glenn, *Sky Storming Yankee* by Clara Studer, published in 1937 (almost seven years after his death), would also promote many of these stories. Unfortunately, the author offered no documentation or reference as to where these origi-nated, other than a list of Curtiss friends and associates she thanked in the acknowledgments for their contri-butions. Oddly, she never mentions any of the Curtiss family. In turn, many of these anecdotes found their way into later works as fact. Sadly, even in an interview with Frank P. Stockbridge in April 1927 for *Popular Science*, Glenn's memory of people and dates was very vague or he did not remember correctly, compounding the issue.

To this day, it is unknown exactly who or what his initial motivation was for all that he accomplished, other than the plain love of creating something new and better.

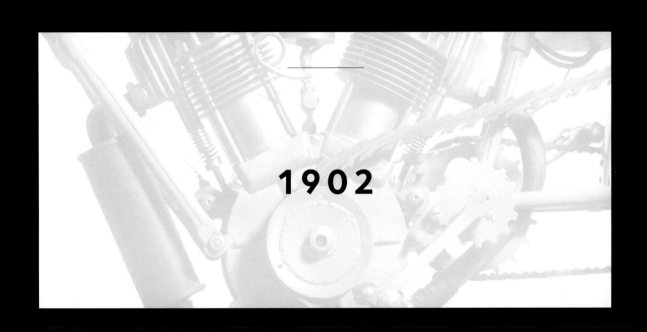

CHAPTER 4

1902

After his son Carlton's death in early February, Glenn devoted more time to his business to help deal with his grief. That spring, having sold his second, improved machine to a Pennsylvania man, he built a third motorcycle, a tandem, and promptly sold it to a commercial baker by the name of Walters in Corning, New York. An interesting note is that Curtiss placed a classified ad in the February 13 issue of *Bicycling World and Motorcycle Review* to sell a maroon-and-black cycle fitted with a Thomas engine and a 1902 Morrow coaster brake for $85. It is not clear where this cycle came from or how it fit into Glenn's experiments or business, other than he had a habit of purchasing other machines, when possible, to study them.

Encouraged by a number of inquiries, his first motorcycle sales, and an order from an unnamed customer in Bergen, New Jersey,[1] Glenn went public that July, forming the G. H. Curtiss Manufacturing Company. Motor oil was now becoming his lifeblood. Shortly afterward, he opened a third store in Corning,[2] intending to use it strictly as a retail motorcycle shop. Always the innovator, Glenn would have a relatively new communication system, telephones, installed in his home and the Hammondsport shop to aid the growth of the business.[3] While Curtiss staffed the Hammondsport shop with the assistance of William "Will" Damoth and E. P. Bauter, the Bath store was run by F. A. Smith, and the Corning store was managed by Flint Barney Feagles. All were long-time friends and bicycle-racing companions. During this time, he would also periodically have a couple of men, John Osborne and Claude Miller, help with the assembling of cycles whenever parts began arriving. Unfortunately, early employment records of the shop or additional information on the Corning retail store are sketchy.

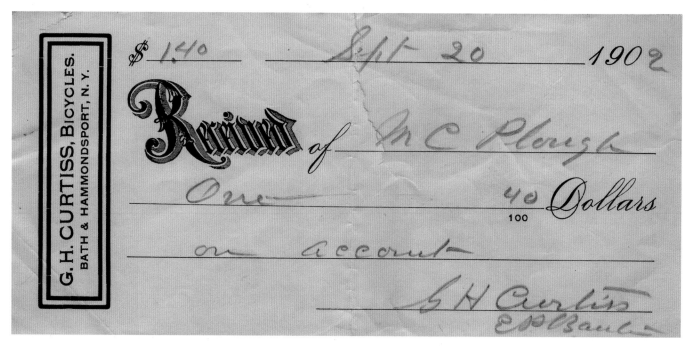

An early shop receipt for the G. H. Curtiss Mfg. Co., 1902

Glenn had contracted with the Kirkham Shop to produce complete engines and with a fabricator in Addison, New York, possibly the Empire Motor Cycle Company, a.k.a. Reliance Motor Cycle Company (producers of the Reliance motorcycle), to produce the frames. He and his men would assemble the finished product. To ensure he had room to go full time into production, Glenn quickly disposed of his sideline horse harness and sewing-machine businesses in October, turning them over to E. P. Bauter.[4]

In his first advertisements of the year, Glenn concentrated on selling engines but also offered for sale a complete motorcycle, and, if one wanted, castings with drawings and assembly instructions to those who wished to build one of their own. He and Kirkham improved the engine the following year by switching the ball bearings to roller bearings in the aluminum casing for a smoother operation.[5]

Looking for a brand name that would be as strong as the machines he built, Glenn would settle on the name Hercules both for his motorcycles and engines. Legend has it that he first produced bicycles under this name, which to date has never been proven. Examination of every advertisement Glenn placed in the two local newspapers, the *Hammondsport Herald* and *Bath Plain Dealer*, shows no mention of his own brand, only those bicycles he was an agent for. Unlike other manufacturers that prominently emblazoned brand names on their machines, Curtiss was low key. "Hercules" could be found only on the medallion or "head badge" on the front column of the motorcycle and sometimes embossed on the motor casing.

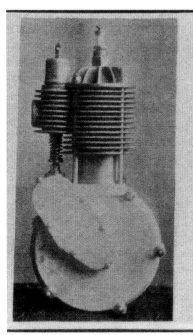

The Hercules

2½ H. P. ball bearing motor for bicycles, tandems, etc. Patents applied for. Just what you have been looking for. Height, 16½ inches; width over all, 4¾ inches; crank case, 8 inches; weight, 35 lbs.; speed, 300 to 2400 revolutions per minute; cylinder, 3 x 3 inches. There are 72 quarter-inch balls in crank shaft bearings; they can never heat or get to cutting, and have only 1-30 the friction of ordinary bearings. The **HERCULES** is the **lightest** and **most powerful** motor of its kind on the market. We sell motors, castings, drawings, accessories and complete motor cycles fitted with these motors. **Tandems** a specialty. Get our prices; they are right.

G. H. CURTISS, SOLE AGENT
HAMMONDSPORT, N. Y.

The first advertisement taken out by Curtiss for the Hercules engine and motorcycle, 1902

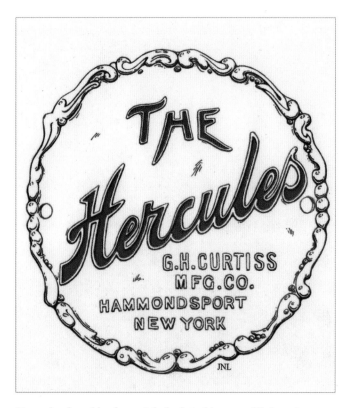

Hercules head badge, nickel-plated brass with hand painted logo in red. *Illustration: Joshua Leisenring*

This initial success and desire to stay in competitive sports prompted Glenn to try his hand at racing his machines. His first endeavor on a one-cylinder Hercules motorcycle took place on Labor Day, September 1, on Coney Island Boulevard in Brooklyn, New York, which was sponsored by the fledgling New York Motor Cycle Club. Interest was so high in the community that A. W. Stratton of the *Hammondsport Herald* accompanied him as a witness. Glenn won a third-place medal against fifteen contestants in the 10-mile Handicap Race, and a second-place trophy, known as the C. F. Splitdorf Cup, for overall time.[6] As an unheard-of newcomer to the scene, this placing garnered him more than a curious notice from the racing community. By the end of the year, it is estimated he had sold five machines under the Hercules brand name.

During his early experimentations, Glenn learned two crucial points that made his motorcycles successful. First was the weight-to-speed ratio. Second was lowering the center of gravity on his machines, which substantially increased efficiency and speed.

On a newsworthy sidenote, in late November, Glenn and Lena would be the first residents to own an automobile in Hammondsport, when he took in trade a used steam-powered car of unknown make for one of his Hercules motorcycles.[7]

Some of Glenn's trophies won while racing motorcycles. The Splitdorf Cup won in 1902 is at the lower left of the image.

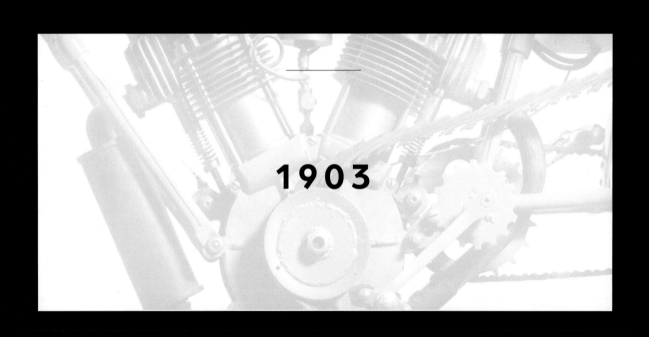

CHAPTER 5

1903

Glenn, now totally absorbed in motorcycle production and racing, dropped sales of bicycles and other sideline interests completely. This included the closing of his retail shops in Bath and the short-lived Corning location by April 1.[1] Oddly, the *Cycle and Automobile Trade Journal* reported on January 1 that Glenn had turned over his cycle motor branch of his business to Charles Kirkham and would manufacture only complete motorcycles. This may be a misunderstanding of Glenn having contracted with Kirkham to build the complete engines while he concentrated on assembling the motorcycles. The Hammondsport shop on the village square was now a manufacturing facility, being retained at least through 1905 as the company grew, and it finally was relocated on the hill behind the Curtiss homestead, overlooking the village. It was also used to handle some bicycle repair, which he felt obligated to do for his early customers, and to sell off remaining stock, take motorcycle orders, etc.[2]

As a result of the Brooklyn race the previous year, a gentleman reported by the *Hammondsport Herald* only as "Mr. Bendix," a contestant in that race, came to visit Glenn a second time in early April (his first visit was in July 1902) to acquire a Hercules engine. His intentions were to use it on a cycle of his own design, with the possibility of purchasing a quantity of them for commercial production.[3] This visit—as well as a previous one from the Industrial Machine Co. of Syracuse, New York, makers of the De Long motorcycle[4]—is significant since it made it apparent that Glenn had a product that would extend far beyond the capabilities of a simple backroom business.

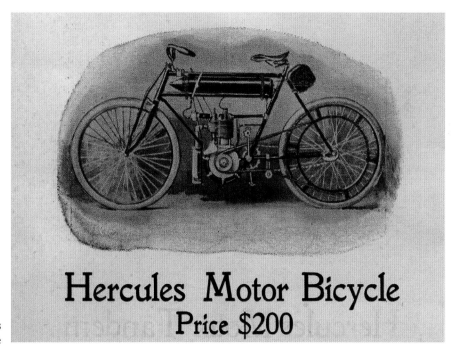

Hercules Motor Bicycle
Price $200

The first commercial Hercules motorcycle

Contemporaneous New York City newspapers reporting on the race of 1902 listed the man as "Hugo Bendix of the New York Motor Cycle Club," who actually took second to Glenn's third in the 10-mile race. Many feel that this is Vincent Hugo Bendix, later founder of one of the largest automotive and aviation electronics manufacturers in the country.[5] In the past it has been thought that Bendix worked for Glenn designing a "torpedo motorcycle" in 1901. This may have stemmed from an article in the May 1938 *Scientific American* claiming that Bendix visited Glenn, asking for advice on a patent he was working on for a spring frame motorcycle when he was nineteen. On the basis of Bendix's birth date of 1881, this would actually put the meeting in the year 1900, which clearly does not fit the Curtiss timeline.

Curtiss's attendance at the New York Automobile Show as a manufacturer, followed by issuing his first catalog in late February, resulted in a sale from California, with inquiries from as far as New Zealand.[6] As a result of this initial publicity, Glenn would receive the first special-delivery registered letter ever to arrive at the village post office.[7] By the end of the following year, the international inquiries would also include Mexico, Honduras, India, Cuba, South Africa, Italy, and Australia.[8]

The year was proving to be very successful for Glenn. While overwhelmed, the shop reportedly produced and shipped twenty cycles as well as an unknown number of engines in the first five months.[9] Glenn immediately set about making major improvements to the cycle and frame and offered the buyer a choice of a black or carmine-red enameled finish.[10]

1903 G. H. Curtiss Mfg. Co. catalog (front)

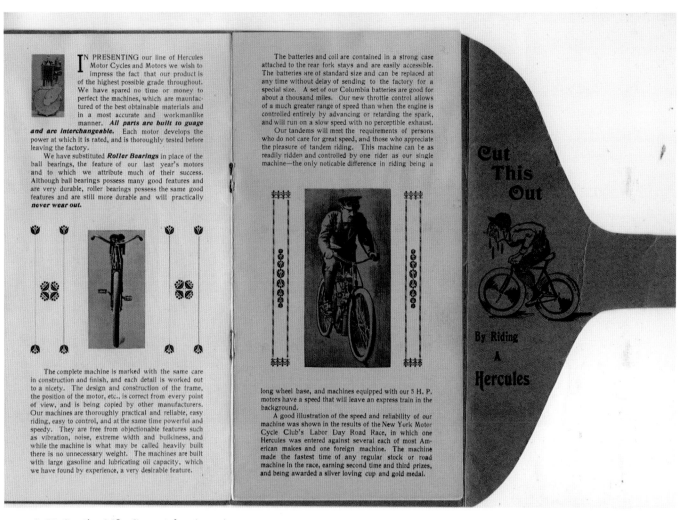

IN PRESENTING our line of Hercules Motor Cycles and Motors we wish to impress the fact that our product is of the highest possible grade throughout. We have spared no time or money to perfect the machines, which are maunfactured of the best obtainable materials and in a most accurate and workmanlike manner. *All parts are built to guage and are interchangeable.* Each motor develops the power at which it is rated, and is thoroughly tested before leaving the factory.

We have substituted *Roller Bearings* in place of the ball bearings, the feature of our last year's motors and to which we attribute much of their success. Although ball bearings possess many good features and are very durable, roller bearings possess the same good features and are still more durable and will practically *never wear out.*

The complete machine is marked with the same care in construction and finish, and each detail is worked out to a nicety. The design and construction of the frame, the position of the motor, etc., is correct from every point of view, and is being copied by other manufacturers. Our machines are thoroughly practical and reliable, easy riding, easy to control, and at the same time powerful and speedy. They are free from objectionable features such as vibration, noise, extreme width and bulkiness, and while the machine is what may be called heavily built there is no unnecessary weight. The machines are built with large gasoline and lubricating oil capacity, which we have found by experience, a very desirable feature.

The batteries and coil are contained in a strong case attached to the rear fork stays and are easily accessible. The batteries are of standard size and can be replaced at any time without delay of sending to the factory for a special size. A set of our Columbia batteries are good for about a thousand miles. Our new throttle control allows of a much greater range of speed than when the engine is controlled entirely by advancing or retarding the spark, and will run on a slow speed with no perceptible exhaust.

Our tandems will meet the requirements of persons who do not care for great speed, and those who appreciate the pleasure of tandem riding. This machine can be as readily ridden and controlled by one rider as our single machine—the only noticable difference in riding being a long wheel base, and machines equipped with our 5 H. P. motors have a speed that will leave an express train in the background.

A good illustration of the speed and reliability of our machine was shown in the results of the New York Motor Cycle Club's Labor Day Road Race, in which one Hercules was entered against several each of most American makes and one foreign machine. The machine made the fastest time of any regular stock or road machine in the race, earning second time and third prizes, and being awarded a silver loving cup and gold medal.

Cut This Out By Riding A Hercules

1903 G. H. Curtiss Mfg. Co. catalog (open)

With Glenn's success as a local businessman also came an expected civic duty from the community. Glenn was appointed a member of the Steuben County Side Path Commissioners representing Hammondsport in March, a position he would hold for the next four years. His duties as commissioner included overseeing the maintenance of bicycle and motorcycle paths throughout the area and the sales of annual bicycle license tags.[11]

As if motorcycles were not enough, Glenn took on automobile agencies as a sideline. Anything propelled by an engine seemed to fascinate him. The first was the Orient Buckboard, produced by Waltham Manufacturing Company in Massachusetts.[12] The Buckboard was a simple small automobile utilizing a one-cylinder engine similar to what Curtiss was producing. Not having space for a showroom, he purchased a Buckboard of his own to demonstrate, and he used the Waltham Company catalogs in a mail-order-type business through his shop, promptly selling two of these to local businessmen. A third Buckboard would be sold to Tank Waters and shipped to Buffalo, New York.

To prove the durability of the one-cylinder-engine-powered machine, Glenn put it through a couple of tests. First, he drove the Buckboard up the steepest hill in the area, Mount Washington, while carrying Mr. Lynn D. Masson, his first customer. This was followed by a 36-mile long-distance trip to Prattsburgh, New York, with Dr. P. L. Alden, his second customer, riding along.[13] Both trips were completed successfully without any mechanical problems.

Glenn would go on to hold automobile franchises for Ford, Thayer-Miller Touring Car Co., and Stoddard Dayton Motorcar Co. by 1908, also using catalogs to sell the vehicles through his shop. His love of automobiles would lead him to personally own a variety of vehicles over the years. Aside from the steam car and Orient Buckboard, the list included a 1907 Ford, 1908 Peerless 4 Cylinder Auto, 1908 Stoddard Dayton Runabout, 1909 Electric Runabout, 1910 EMF Touring Car, 1911 Winton Touring Car, 1912 Flanders Electric, 1912 Chalmers Four, 1913 Jeffery Six, 1913 Keeton, 1914 Buick Touring Car, 1914 Detroit Electric Roadster, 1915 Pierce Arrow Touring Car, and 1917 Smith Motor Wheel, as well as four custom-built autos by Brunn & Co. of Buffalo, New York, which were made between 1912 and 1916. Glenn, however, never held on to any one vehicle very long, either selling, trading, or, in a couple of instances, giving them away as gifts.

While having briefly promoted them only as "a specialty" in the 1902 advertisements, Glenn would strongly emphasize the tandem motorcycle in the lineup this year. The main sales point was that a lady's skirt would not be torn or soiled during a ride.[14] Glenn saw the motorcycle as a family-related sport. Sadly, the following year's catalog would announce that "as these machines are not meeting with their deserved popularity, we have decided not to catalog them . . . but will build them to order." A conversion kit would finally replace the custom-built tandem in 1906.[15]

The manufacturing end of the business quickly outgrew the small backroom of his Hammondsport village shop. It was immediately expanded to include the large barn in the back, with an expansion added to its side.[16] After less than a full year in business, the shop now employed up to seven men on a regular basis.[17] Lena Curtiss would join the business as the office manager to allow her husband to do what he loved best—build and invent.

L. D. Masson seated in his new Orient Buckboard purchased from Glenn H. Curtiss, 1903

Glenn H. Curtiss in his 1908 Peerless four-cylinder automobile

A Hercules Tandem with Mr. and Mrs. C. L. "Tank" Waters, taken in November 1903 and featuring the new Hercules V-Twin engine

A Hercules Tandem motorcycle as pictured in the 1903 catalog

Hercules Motor Tandem

Family members also believed that Lena's actions were designed to help with her own grief. On April 28, she lost her father, Guy Neff, after a long bout with stomach cancer. This was followed by the loss of Glenn's beloved grandmother and matriarch of the family, Ruth Curtiss, on September 24. Lena had grown close to Ruth and was her primary caregiver. Since Lena was an only child, her mother moved in with the Curtisses for a short time after her father's death.

Since G. H. (the name he preferred his friends call him) considered his first racing event a poor showing, he set out to develop a more powerful engine, again seeking the help of Charles Kirkham. Glenn's new creation, the V-Twin, a two-cylinder, air-cooled, 5hp engine, was the first of its kind in the United States. The engine, which powered his new Hercules, was publicly made available in April. The machine made its debut in New York City at the New York City Motorcycle Club Riverdale Hill Climb on May 30. He easily took first place out of the twenty entrants, completing the half-mile climb in fifty-one seconds and winning a gold medal. Heading directly to Yonkers, a few miles away, Glenn won the National Cycling Association Championship medal for the 1-mile race and a first-place medal for the 5-mile race at the Empire City Race Track, thereby vindicating himself for his personally perceived loss the year before. More astonishing was the fact that he set a world speed record of little over one minute for the 1-mile race. To Glenn's embarrassment, he was hailed as a hero, with a parade and a serenade by the local band, and was presented a cut-glass bowl by the community on his return to Hammondsport.[18]

The 1903 Hill Climb Medal won by Curtiss. Reverse is engraved "HILL CLIMBING TEST DECORATION DAY 1903 WON BY G. H. CURTISS." *Courtesy of National Air & Space Museum, Smithsonian; photo by Richard Leisenring Jr.*

Glenn's triple win in the New York metropolitan-area events was enough to convince a few of his racing competitors to switch over to his machines. One such convert was a well-known amateur racer, William E. Wahrenberger of New York City, who had competed against Glenn in New York while using an Orient cycle. Immediately acquiring one of Glenn's Hercules V-Twin motorcycles, Wahrenberger would use it in the following July 4 Endurance Run from New York City to Worcester, Massachusetts. This is the first recorded use of a Hercules by someone other than Glenn in a major event.[19]

Looking to continue his winning streak, Glenn, with Charles Kirkham as partner, entered the races scheduled in Brooklyn on September 5 through 7, but currently, there is no evidence they attended. However,

Glenn did participate in a 2-mile race at the New York State Fair in Syracuse, New York, on September 12. While in the lead by a considerable distance, the sparking device fell off and, as reported, "his machine stopped dead still. Mr. Curtiss had the mortification of seeing his competitors pass one by one. . . . The prize was a silver cup[,] which he would have given considerable to own."[20] During the meet, Glenn would join the newly formed Federation of American Motorcyclists (FAM) as a charter member.

Glenn's V-Twin immediately became the focus of the industry, with other companies looking either to add his engines to their cycles or develop their own version in order to compete. One such company was the Harry Geer Co. of St. Louis, Missouri, which in later 1904 through 1909 used Hercules/Curtiss V-Twin-model engines for use on their cycles sold under the name Green Egg.[21] Indian Motorcycles, a major competitor of Glenn's, commercially offered their first two-cylinder engine in 1906, followed by Harley-Davidson in 1909.

Glenn's high profile in the new motorcycle industry due to his fast machines and racing prowess gave him his first brush with the fledgling aeronautical community. It came as a special order for a V-4, four-cylinder engine, rated at 10 hp, and the first of many variations for Curtiss. The customer who requested this development was amateur balloonist Thomas Benbow, representing a group calling themselves the American Aerial Navigation Company, for use on their dirigible the "Montana Meteor." The craft was being built at the Prof. Carl Myers Balloon Farm in Frankfort, New York. Benbow had heard of Hercules engines through Myers, who had earlier purchased an engine from Glenn. This

Benbow project would also be newsworthy on another level. Glenn, with the help of Charles Kirkham, personally delivered the engine that October to the balloon farm. Traveling from Hammondsport to Frankfort with the new engine strapped to the back of Glenn's motorcycle,[22] the pair easily made the round trip of over 300 miles without incident on their motorcycles, quite a feat considering only dirt and secondary roads existed between the two points at the time.

His second brush with these aeronauts would arrive in the form of a telegram from a G. A. Faulkner, an Oakland, California, motorcycle dealer who, in late July 1904, ordered five 5hp Hercules V-Twin motorcycle engines COD—the largest "cash on delivery" order Glenn would receive. One of these engines was reported "to be used in a flying machine." The order was for Captain Thomas Scott Baldwin, a celebrity aeronaut of the day. He had hopes of using it on his dirigible the "California Arrow."[23] Both Benbow and Baldwin had set their sights on a large cash prize of an unheard-of $100,000 being offered for the first dirigible to make a motorized flight around a selected course at the upcoming 1904 Louisiana Purchase Exposition in St. Louis, Missouri. Each looked to Curtiss-built engines to help them achieve their goal. This would also be the start of inquiries and visits from various other experimenters, each one trying to perfect the art of manned flight with their newfangled machines over the next few years.

On the home front, Glenn's sister, Rutha, would graduate from the Western New York Institution for Deaf Mutes, and she came to stay at the Curtiss homestead temporarily until she started the next school season as a teacher at the institute.

Montana Meteor

Thomas Benbow's "Montana Meteor"

Myers Balloon Farm in Frankfort, New York

One of Thomas S. Baldwin's first Hercules V-Twin engines set up for his dirigible

A Hercules one-cylinder engine set up for use on a dirigible

CHAPTER 6

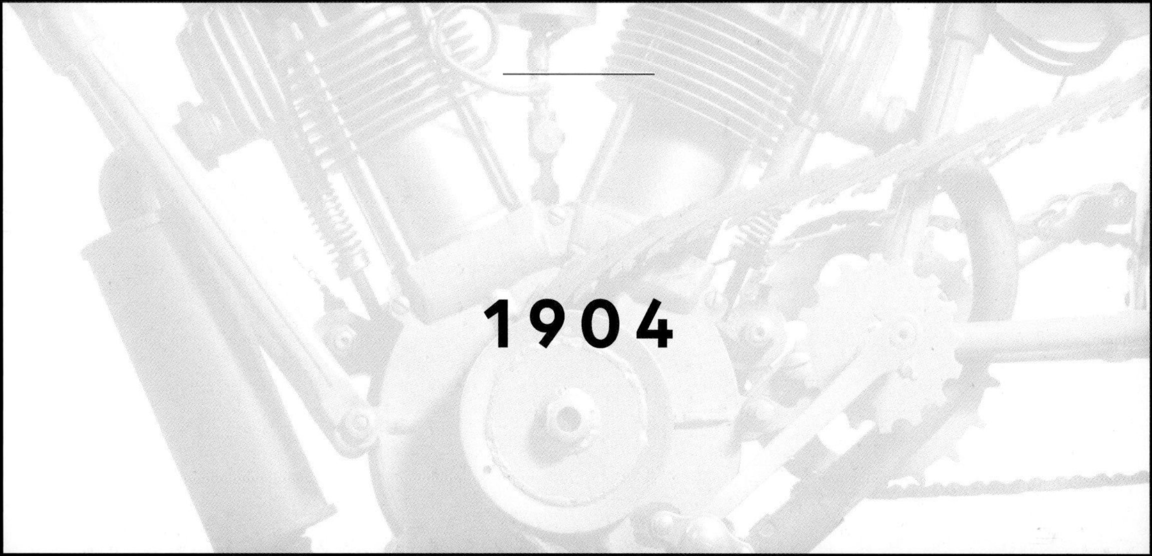

1904

In January, shortly after attending the New York Auto Show, Glenn headed to Ormond Beach, Florida, with his special V-Twin Hercules Racer motorcycle to enter the newly formed speed trials known as the Carnival of Speed. While there, Glenn was the guest of G. T. Spaulding. Undaunted by the competition, Glenn won the 1-mile race trophy on January 29 in fifty-nine seconds, breaking all previous American records, and the 10-mile race trophy the following day in eight minutes and fifty-four seconds, setting another world record in his early racing career.[1] It was a record that would stand for several years. Oscar Hedstrom, on an Indian motorcycle—the man everyone thought would win—came in second. As a reward for his employees, Glenn shipped back to the plant a crate of oranges, an expensive and seasonal treat not commonly available in a small rural New York village in the winter of 1904. A few months later, one of the promoters of the Ormond Beach event, W. J. Morgan, visited the Curtiss facilities. Surprised at how small it was, Morgan publicly urged the community in an open letter published in the *Hammondsport Herald* to do all they could to help the company expand.[2]

Glenn H. Curtiss on his Hercules V-Twin racer, which set a record at Ormond Beach in January 1904

1904 Hercules Twin Racer. A duplicate of that used by Curtiss at Ormond Beach. *Courtesy of Dale Axelrod, photo by Derek Snyder*

Color postcard of trophies won by Glenn H. Curtiss at Ormond Beach, 1904

Glenn would revisit the Memorial Day New York Riverdale Hill Climb that May. Finding his 5hp V-Twin barred by the new 3½hp limit, he took second place with a one-cylinder Hercules machine even while hampered with fuel mixture problems.[3] He missed out on first place by 2.8 seconds.

Glenn's next shot at competition would not fare as well. Over the Fourth of July weekend, Glenn traveled downstate to a series of races starting in New York City and culminating in Cambridge, Maryland. Unfortunately, due to two separate minor mishaps while on the courses, he was unable to capture any of the coveted prizes. Nonetheless, his machine received praises from a number of his competitors and fellow sportsmen.[4]

A week after the event, E. H. Corson, an executive from Hendee Manufacturing Company, makers of the Indian motorcycle, visited the Curtiss shop in Hammondsport to get a better understanding of this new competition winning out over their motorcycles. Amazed at what he saw after finding the entire manufacturing operation located in this tiny facility on the village square, Corson was quoted as saying in a bemused way, "This is it?!" Obviously, like W. J. Morgan, he was expecting something on a larger scale.[5]

Meanwhile, not everyone in the area seemed to be excited about the motorcycle becoming a popular mode of transportation. The Village of Bath would periodically, without prior notice, change the ordinances regarding the use of the bicycle cinder paths in regard to the motorcycles. The *Hammondsport Herald* called the changes "the vicious Bath ordinances." It seemed that one week the motorcycles were limited to using only the bicycle paths, while on another week only the roads, causing a great deal of confusion to motorcyclists. As a result, Will Damoth was arrested and taken to jail to pay a fine of $3.00 for using the cinder path, while a few days later,

Charles Kirkham was detained by Bath police for using the road.[6] This animosity seemed to continue on and off for quite some time, since, two years later, Albert D. Cook was arrested for driving on the bicycle path on a Sunday. Fortunately for Cook, Glenn and several friends were test-driving an automobile in the area and had witnessed the incident. Intervening, Glenn assured the arresting officer that Cook would appear in court on Monday. He was allowed to proceed without going to jail.[7]

According to the *Buffalo Courier* on September 3, Glenn Curtiss was to make a special appearance at the Tonawanda Driving Park in Tonawanda, New York, with the intention of breaking his record for the 10-mile run he set back in January. However, no mention was ever made regarding this new record attempt after he and Charles Kirkham competed in the Labor Day event on September 5. During the races, Glenn took second in the 15 Mile Open and first in the Five Mile Open and five Mile Exhibition races. Kirkham took second in the Five Mile Open. Both rode Hercules cycles.

On October 21, it was reported in the *Hammondsport Herald* that Glenn was about the village on crutches as a result of blood poisoning in his foot. Sadly, no explanation or follow-up regarding his health was ever given. However, it did not seem serious enough to keep him from registering as a participant in the October races scheduled to be held in Buffalo later that month. Ironically, the event was canceled due to bad weather.

Glenn would attend another motorcycle race in Hackensack, New Jersey, later on November 8.[8] Known as the Election Day Races, the event was sponsored by the newly formed North Jersey Motor Cycle Club and held at the Bogota Driving Track.[9] Oddly, it is not clear if he actually went to participate. It is believed that he attended the event strictly as a spectator in

Hercules Model 24 5hp, two-cylinder engine. The 24 stood for two-cylinder, 1904. *Courtesy Airpower Museum, Ottumwa, Iowa*

1903/1904 Hercules single-cylinder Hercules motorcycle. *Photo: Mark Preston*

G. H. Curtiss Mfg. Co. building located next to the Curtiss homestead. Note the Wind Wagon in the right back, 1906.

order to see the new Indian V-Twin being developed by Oscar Hedstrom and rumored to make a showing at the event. Hedstrom, chief exhibition racer and cofounder of Indian motorcycles, had lost to Glenn at Ormond Beach back in January, as he would on other occasions, and was determined to outclass him.

At this time, other entities, such as Owego, New York's Board of Trade and later the city of Rochester, began approaching Glenn with various offers to relocate his thriving business there.[10] Even the Reliance Motorcycle Works, Glenn's early supplier of cycle frames, approached him with an offer of merging and relocating to Owego.[11] These overtures caused much concern among the locals, but the young native assured them his loyalty was to Hammondsport. As proof, Glenn would relocate production of his cycles to the Curtiss homestead off Pulteney Street and build a new 20-by-60-foot, two-story plant building next to the house in September. [12]

Business at the Curtiss plant also greatly benefited the Kirkham machine shop, since they were now employing fifteen men and outgrowing their current location just to meet Glenn's needs. Earlier in the previous year, Charles began designing his own motorcycle and would deliver his first sale to a customer in Dayton, Ohio, in May. He had hopes of going into commercial production.[13] After cutting ties with Curtiss later in 1905, he formed the Kirkham Motor Manufacturing Company, also known as the Kirkham Motor Cycle Company, and relocated to the village of Bath, New York. Charles would experiment with motorcycle designs, produce a few Curtiss/Kirkham-type motorcycle engines, develop and commercially produce automobile engines, and, much later, try his hand at aviation.[14] In the spirit of friendly competition, he would occasionally race a "Kirkham cycle" against Glenn between 1904 and 1907. Charles would eventually work for the Curtiss Aeroplane & Motor Company from 1915 to 1920 and briefly for the Curtiss Aerocar Company. Forming an aviation-consulting business in 1920, Kirkham would finally retire in 1940, passing away in 1969.

A significant change in the branding of the motorcycles would come in the fall. After Curtiss was notified by another company that they held the rights to the brand name Hercules, all models thereafter were branded under the Curtiss name.[15] Following suit with the script Hercules logo, the now-famous Curtiss logo was designed for Glenn by a local Hammondsport barber and friend Claude C. Jenkins. The new logo would be conspicuously displayed on the sides of the cycle's gas tank. It would also replace the Hercules name on the engine casing as well as the head badge located on the front column.

Since high quality was always foremost in Glenn's mind, the 1904 catalog would promote the high standard of finish now found on all Curtiss-built motorcycles. The enamel paint was applied in four layers. All bright metal parts except aluminum were nickel-plated. At this point, the standard color offered up to 1908 was black. Other colors could be special ordered, but as pointed out in the catalog, this would delay the delivery of the machine. It was also suggested for the first time that the V-Twin engines were suitable for automobiles and airships.[16] While not advertised by Curtiss, several one-cylinder engines were converted for use as outboard motors on small boats.

Curtiss head badge, nickel-plated brass with hand-painted logo in red. *Illustration: Joshua Leisenring*

Another unique feature was the introduction of the twist grip on the left side of the handlebar to control the switch and spark advance. The throttle was a lever attached to the left-hand side of the machine.[17]

Starting with the first controlled circular flight in August by Lincoln Beachey in Captain Baldwin's "California Arrow," followed by the historic public flight at the 1904 Louisiana Purchase Exposition on October 25, which was piloted by Roy Knabenshue, the demand for Glenn's engines to be used in other types of transportation was on a steady rise. Team Baldwin's feat at the expo would earn Glenn high recognition specifically for his engine (it is unknown if a medal was actually issued).

Ever the flamboyant showman, Baldwin told a story in 1909 of how, after seeing a Curtiss motor, he "hurried across the continent to Hammondsport. He found young Curtiss in his shop and that same day gave an order for an airship motor."[18] In later years, he would fabricate and embellish several versions of a story of how he had to chase Glenn down and personally visit him in order to obtain his engine—a used one at that—when in fact the two did not meet until after Baldwin's accomplishment in St. Louis. The engine that Baldwin used had been previously purchased through an agent in California. On November 20, 1904, Baldwin appeared unannounced in Hammondsport for the first time to personally meet and congratulate Glenn on his part in the achievement, as well as give the *Hammondsport Herald* an exclusive interview. The success of the engine as used in a dirigible came as a complete surprise to Glenn, since he had heard nothing of it until then.[19]

While Glenn was quickly developing a close relationship with Baldwin, it has been said that he initially did not take these aeronauts seriously. Regardless, he did realize early on that their patronage was important and a good business venture. Listening closely to the needs of these pioneers, Glenn realized that he could tweak the engines to lighten them and gain more horsepower to meet their requirements. To enhance sales, the *1905 Curtiss Motorcycle and Engine* catalog prominently features an endorsement from T. S. Baldwin and the Baldwin Airship Company of San Francisco, California.

A curious yet unverified quote regarding these aeronauts has Glenn saying, "I get twice as much money for my motors from those aviation cranks" (the term "crank" meant a person with strange ideas). This seems to imply that he was knowingly taking advantage of them. A quote of this nature (which Clara Studer, in *Sky Storming Yankee*, indicates originated in an interview with Ritchie Betts, editor of the *Bicycling World and Motorcycle Review*) is generally out of character for Glenn and his business ethics. While his first sales were of motorcycle engines at a flat price regardless of who bought them, an examination of the catalogs for the period shows that a 5hp V-Twin motorcycle engine by itself, without any accessories, sold for $150. The airship version offered shortly after, in 1906, was rated at a more powerful 7 hp, was lighter in weight, and came with the engine mounted in a wooden frame with aluminum braces, balance wheel, starting crank, coils, carburetor, three batteries, exhaust, oiling device, extra spark plugs, wrench, sprockets, two sets of roller bearers, etc. for $260. More interesting is that historian Geoffrey Stein, in his book *The Motorcycle Industry in New York State*, found that an unidentified aviator of the period was actually publicly quoted in print as saying, "Curtiss could get twice as much money for an aeronautical engine as for a complete motorcycle."[20]

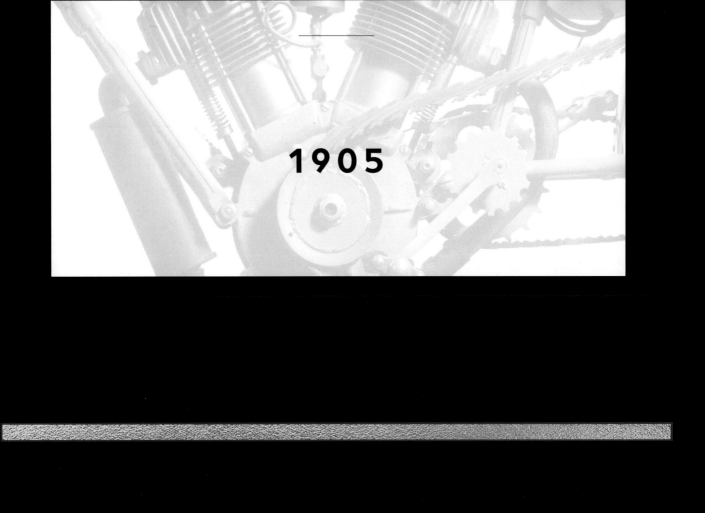

1905

Glenn's absence from the shop was becoming quite conspicuous with the locals, since he was spending a large amount of his time traveling to shows, meeting with new distributors, exhibitions, and races. Attending the January Auto Show in New York with the intention of participating in the speed trials a week later at Ormond Beach, Glenn had something out of the ordinary in mind. Earlier in the winter, he had his employees mount a V-4 on a motorcycle to race there.

Sadly, the cycle was never put to the test, since Glenn was unable to make it to Ormond due to a major delay in dealing with the overwhelming number of orders and inquiries received at the show. While promoting the success of Baldwin's dirigible at St. Louis with a Hercules (now Curtiss) engine, he would manage to place an agency with a well-known dealer in England.[1] While it was a high point for the business, the dealer was never named.

Glenn H. Curtiss posed on his V-4 racing cycle, intended to run at Ormond Beach, Florida, in January 1905 and again on July 4, 1908, in Buffalo, New York.

The experimental V-4 engine (originally designed for Benbow in 1903) was unique, since it actually consisted of two separate V-Twins set side by side and connected with a central driveshaft, one engine running in reverse of the other. This particular V-4 was made up of two slightly altered 5hp Curtiss Model 24s, giving a rating of 10 hp. The drive gear was located on the shaft between the engines and was chain-driven instead of by a belt, as commonly used on his other cycles. The chain drive was a feature he never used again. The engine, while workable, was not practical in a commercial sense and was never offered for retail sale. As for the motorcycle, it is not known if Glenn ever tried it out personally to test its potential speed or what finally became of the machine. However, three years later, the cycle would briefly reappear, at least in print. The *Buffalo Courier* of Buffalo, New York, reported on June 28 and again on July 4, 1908, that Curtiss was sending the V-4 cycle to the July 4 event at Kenilworth Park, in an attempt to set new speed records. The advertised goal was a mile in thirty seconds and 10 miles in eight minutes. The event was postponed to the next day due to rain, and the Curtiss cycle was withdrawn. At this time, it is not known who would have made the attempt in Glenn's absence, since he was personally tied up at another demonstration that day.

The Model 110 Racing Motorcycle

The Curtiss homestead would see two additions to the family this year. Glenn's mother, Lua, divorced her second husband, Charles Adams, and moved back to Hammondsport with a new member of the family, Glenn's half brother, eight-year-old G. Carl Adams. Glenn would take charge of Carl and treated him as if he were his own son.[2]

Something unique added to the 1905 annual catalog. While he was custom-building racing motorcycles in past years, this is the only time Curtiss offered a V-Twin

1904 Curtiss 4hp, single-cylinder motorcycle engine

racing cycle as a stock item. Known as the "Curtiss 110," it was streamlined to weigh only 110 pounds and utilized a special 6hp, two-cylinder model 25 engine with no exhaust pipe or unnecessary fittings. The shorter wheelbase, as warned, made it unsatisfactory for road use. As opposed to the standard cycle, the 110 used a special set of batteries contained in a tube placed vertically along the rear seat post.

Since many of his racing cycles and experimental machines, such as the V-4 or 110, were built during the fall and winter months, road testing them was not possible due to the bad weather and amount of snowfall the area received. However, Glenn and his employees did periodically try out his cycles during the summer months at the Stony Brook Farm racetrack, a couple of miles down the road from his home, when not running around the village and surrounding area. Built as a horse track, it also doubled as a bicycle track for the local bicycle-racing team, with which Glenn was extremely familiar. The track may have been the site of his early motorcycle experiments back in 1900 and 1901.

Always hands on and a firm believer in his products, Glenn rode his personal motorcycle to Albany, New York, in late April to fix a customer's cycle. Leaving on a Tuesday night and returning by Thursday morning, Glenn completed the roughly 350-mile round trip and repair job in record time.[3]

One of the more unusual events Glenn attended was in Chicago, Illinois, over July 4 as a guest of the Chicago Motorcycle Club. He was there not as a competitor but merely to run an exhibition race in order to demonstrate his 5hp V-Twin motorcycle for the general public.[4] Circling the half-mile track ten times in five minutes and five seconds proved to be a crowd pleaser. This publicity stunt drew several thousand spectators.

The G. H. Curtiss Mfg. Co. crew. Lena Curtiss is in the far back, to the left. Glenn is kneeling just behind the front wheel of the motorcycle, with A. D. Cook holding the handlebars.

Lena Curtiss and Tank Waters demonstrating the new Curtiss Side Car

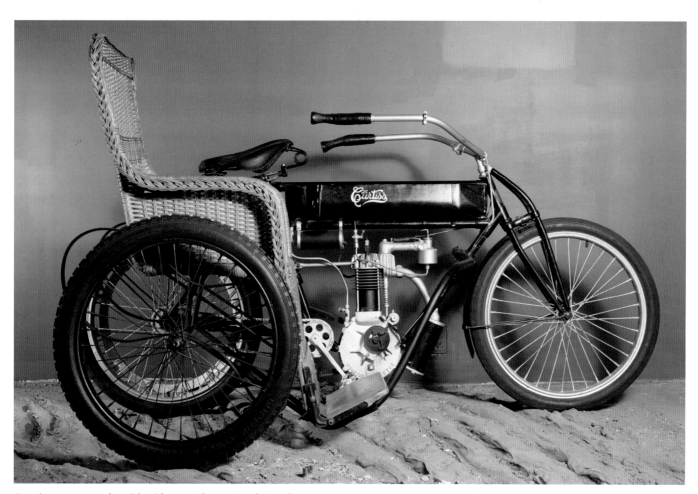

Curtiss motorcycle with sidecar. *Photo: Derek Snyder*

Interior of the Curtiss shop in 1905, with Glenn H. Curtiss standing to the far left. The employees are assembling V-Twin motorcycles. Notice the new Curtiss logo on the cycle in the foreground.

Another interior photo, but with the same employees at the opposite end of the room clowning around with beer bottles and a pipe in defiance of the "No Smoking" sign

The Lewis & Clark medal that Curtiss was awarded. The original medal was engraved "Highest Award / Gasoline Engine / G. H. Curtiss Mfg. Co. / Hammondsport / N.Y." The original was lost many years ago. *Photo: Richard Leisenring Jr.*

Glenn found that the increased demand for his machines, as in previous years, brought the need for more expansions, along with a larger workforce (now up to thirty full- and part-time workers) and an improved product line. In November, another 20-by-50-foot, two-story building would be erected, doubling the size and capacity of the plant.[5] The introduction of a wicker sidecar and a Curtiss-brand spark plug was two of the newest product additions as the company progressed.[6] The spark plug would undergo several improvements and finally be patented in 1909. While items appear listed from time to time in various catalogs as "patent applied for" or "patent pending," currently a search for motorcycle-related patents associated with Curtiss has proved fruitless, and it is possible that while he intended to apply for the patents, Glenn or his company simply never followed through.

Glenn realized that in order to continue on this path of success, he needed more capital. In October, he incorporated the G. H. Curtiss Manufacturing Company, bringing in over ten investors, many of them local business owners. This move brought in a large amount of cash to help secure the company's future. He would now take the position of general manager, with an annual income of $1,800[7] (equivalent of $62,000 in 2023). It was about this time that production and manufacture of cycle parts, originally outsourced to Kirkham and the Addison shops, was transferred entirely to the Curtiss works and fabricated in-house with no outsourcing.

Some of the awards Glenn garnered this year included both the first-place medal and time prize trophy in the 25-mile road race in Waltham, Massachusetts,

on August 8, as part of the FAM Endurance Run from New York City to Waltham. Riding against twenty-six competitors, his V-Twin cycle was put to the test.[8] J. P. Bruyere (an independent racer with no relation to the Curtiss Company) would also win a gold medal for the 250-mile endurance run on a Curtiss cycle, including setting a speed record for the mile run—a welcome boost for the credibility of the Curtiss machines.

On September 18, Glenn took two first-place trophies for 5-mile races in different classes on his single and V-Twin, setting two world records as well as winning the trophy for the 3-mile race for the single-cylinder class at the New York State Fair in Syracuse, New York.[11]

A few months later, when Glenn received his trophies from the state fair, he found that all of them had been engraved "Won by George H. Curtiss." This case of mistaken identity would plague Glenn through most of his early career. Since he would generally register for events as G. H. Curtiss, race officials, news reporters, and a variety of others naturally assumed it stood for the more common name of George.

His only known medal specifically awarded for his aeronautical engines would come at the Lewis & Clark Exposition in Portland, Oregon, in October, again due to Baldwin using a Curtiss engine advertised as a V-Twin (believed by some to be a V-4) on his dirigible the "City of Portland." This would be piloted by Lincoln Beachey, who would later go on to work with Curtiss as an exhibition flier in 1911.[10] An interesting sidenote to the names of Baldwin's dirigibles, which caused periodic confusion, is that he called all of them "California Arrows" until a name was officially settled on for the individual dirigible.

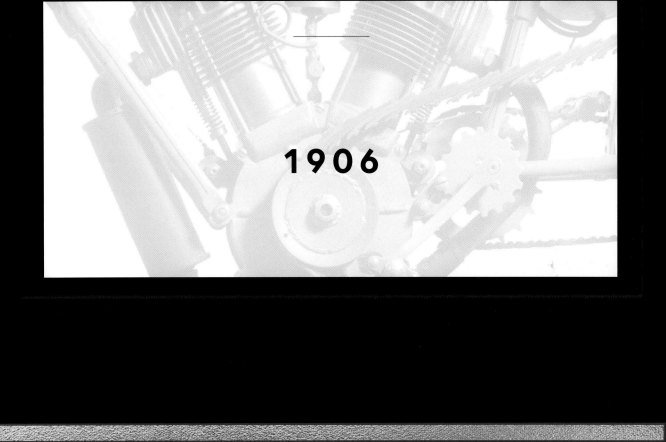

1906

Hammondsport was quickly becoming the center of attention for American aviation experimentation, thanks to the superior Curtiss engines. In response, Glenn would, for the first time, exhibit aeronautical engines at the annual New York Auto Show that January along with his cycles. It was here he briefly met the famed inventor Alexander Graham Bell, who placed an engine order with Curtiss.

Immediately after returning from the show, Glenn realized that the company was drastically falling behind in filling orders, creating a three-month backlog. The crew was already working thirteen-to-fifteen-hour days, and in order to clear the backlog, a night shift was instituted, which almost doubled the workforce overnight.[1] Shortly after, another building addition was constructed in March to house the business office and enameling department.[2]

In March, Tank Waters moved back to Hammondsport from Buffalo with his wife of three years, Elizabeth "Bess" Hopewell. He brought with him his business, the Motorcycle Equipment and Supply Company (MESCO), which he had started in 1904. Glenn invited Waters to use the Curtiss facilities to produce the new Erie brand of motorcycle—introduced in

Erie logo. *Illustration: Joshua Leisenring*

1905—as well as motorcycle kits first produced under the MESCO name by Waters in 1904, until a building on Lake Street was constructed. At this point, Tank's business was also gaining notice from around the world. One order of note would arrive in March from Prince Andrew Gagarin of St. Petersburg, Russia, for an Erie 2hp motorcycle kit.[3] Tank issued an Erie catalog, a MESCO catalog, and an annual motorcycle newspaper titled *The Motorcyclist*. Tank also had a sideline business known as the Buffalo Bicycle Supply Co., which took over Glenn's remaining bicycle merchandise and his dormant agencies.

For Curtiss-brand motorcycles, a new innovation was offered this year—the handlebar twist throttle, which was located on the right side, opposite the switch and spark advance twist grip located on the left.[4] While some historians credit Glenn as the inventor of the twist throttle, Glenn and his company never laid claim to it. The concept originally dates back to the 1860s, with the twist throttle first used on the 1867 Roper Steam Velocipede (an early form of motorcycle). Hendee Motorcycle Company began using a version of it in 1904 on their production-made Indian motorcycles. This attribution may be in confusion with the 1904 twist grip for the switch and spark advance that Glenn introduced.

C. L. and Bess Waters on a one-cylinder tandem, William Damoth on a one-cylinder cycle, and Glenn and Lena Curtiss on a two-cylinder tandem

Glenn kept up with his duties as a Side Path commissioner. On April 11, 1906, he took time out of his busy schedule to oversee the layout of cinders on the bicycle path as far as the Pleasant Valley Wine Company property, 2 miles down the road from the village square. He advertised for bicyclists and motorcyclists to help complete the path as a public service.

On July 4, Glenn attended the Rochester, New York, motorcycle races with a newly designed racing cycle. As a treat for the holiday, he closed the plant and invited his entire workforce to attend the races.[5] The bike easily won the 3-mile contest and the 5-mile national championship. To his chagrin, after winning the 5-mile race, Glenn's machine was disqualified from the event for being slightly heavier than the imposed 110-pound weight limit. His other cycle, a standard V-Twin, ridden by his new racing companion and employee Albert D. Cook, was also disqualified after taking second place for the same reason.[6] Sadly, while making a good showing the Curtiss team went home empty-handed. Glenn would later voice the need for two weight categories to avoid this in the future.

One young man who would race against Glenn and Cook that day was William C. Chadeayne, captain of the Buffalo Motorcycle Club. He walked away with two trophies—the 10-Mile Rochester Handicap and the 3-Mile Flying Start. Hailing from Buffalo, New York, Chadeayne was a well-known regional champion racer, holding several national speed records and one transcontinental US record while representing Thomas Auto-Bi Motorcycles. Auto-Bi was later reorganized as the Greyhound Motor Works of Buffalo in 1909 with Chadeayne as the owner, producing Greyhound motorcycles. By 1915, with the industry extremely competitive, Chadeayne had phased out of the motorcycle business and went to work for the Curtiss Aeroplane & Motor Co., eventually moving to Hammondsport and cofounding the Aerial Service Corporation, which is still in business today as Mercury Corp.

While Glenn's competitive streak continually pushed him to win, it is obvious that he did lose a race on more than one occasion and graciously accepted defeat, using it as a motivation to excel in the next. Glenn was once quoted regarding his competitive nature when it came to racing: "What is the need of racing unless you think you are going to win? And if you are beaten before you start, why take a chance?"[7]

Following the San Francisco earthquake that May, Baldwin relocated the remains of his dirigible company to Hammondsport, taking advantage of the Curtiss facility. Baldwin would also gently pull Glenn into the sport of dirigible flying by inviting him to travel to several events such as the Dayton, Ohio, Air Meet that September, primarily serving as an engine expert. It was here that both men met the Wright brothers briefly during an exhibition flight of Baldwin's dirigible.[8] Glenn would later write to the brothers, offering them an engine free of charge for their experiments, to which they politely turned him down, feeling that theirs would suffice.

Since the Curtiss Company was now supplying Baldwin with engines and lending space for dirigible construction, it seemed only natural to supply the propellers. In order to test out their efficiency, Glenn built a three-wheeled machine for the occasion. Known as the "Wind Wagon," the wood-framed tricycle weighed in at 300 pounds. It was propelled by a 6-foot dirigible propeller powered by a two-cylinder Curtiss motorcycle engine and could attain 30 mph on flat roads. While novel and a great attention-getter, the machine was short lived. Since it scared horses and caused quite a commotion when operated around the village,[9] Glenn quickly agreed not to run it after a few tests, to keep peace with the locals. To compensate, he created an airboat to run on nearby Keuka Lake, followed by an iceboat later that winter. The village would institute a speed limit of 10 mph for all motorized vehicles regardless of power source the following year.[10]

William C. Chadeayne with a Greyhound motorcycle built for the Buffalo Fire Department

Glenn H. Curtiss and Thomas Baldwin posing with the Wind Wagon, 1906

Glenn H. Curtiss on an Ice Sled, powered by a new four-cylinder, in-line engine with a dirigible propeller, 1907

During the first week in December, while Glenn was at the New York Auto Show, he was approached by two separate experimenters, Prof. William H. Pickering of Harvard University and Dr. Julian P. Thomas, a celebrity balloonist of New York City, looking for something other than the current engines offered for their aeronautical experiments.[11] Accepting the challenge, the Curtiss team would create a 20hp, four-cylinder, air-cooled, in-line engine specifically for aeronautical use, which was supplied to both men. Over in the aeronautical display at the show hung Baldwin's "California Arrow" and a dirigible built by Leo Stevens for a Major Miller, both featuring Curtiss engines. Professor Pickering would actually have a Wind Wagon of his own design displayed resembling a large tricycle made of bicycle tubing with a four-bladed propeller in back. A later model was said to be powered by a Curtiss engine. These "wagons" were now a novel mode of transportation.

Excited by the growth of the company and support of the community, Glenn instituted the Suggestion System for his workers with cash awards to help with morale and improve products and productivity. The program proved to be so popular with his employees that several awards were handed out in the first few weeks.[12] From handing out cash rewards to little treats such as crates of oranges or Christmas hams, Glenn understood that rewarding his employees not only was good for the business, but because of his upbringing, knew it was the right thing to do. For years, he would see that the reward system was practiced no matter how large the company got.

Another aeronaut to relocate to Hammondsport to work with Curtiss and later Baldwin was Charles Oliver Jones, with his designs for two different flying machines and a dirigible. In September, Glenn and company would receive an order from Jones for an engine like nothing they had ever built before—an air-cooled, 40hp, eight-cylinder engine, known as a V-8.[13]

A group of at least seven Curtiss motorcycles being shipped to distributer G. A. Faulkner in Oakland, California, from Hammondsport, New York, in 1906

While it is unknown who built the first V-8 engine, the first patent granted for a V-8 engine was in France in 1902 to Léon Levavasseur. Known as the "Antoinette," it appeared in competition speedboats in 1904. However, the Curtiss V-8 is considered the first aeronautical V-8 engine in the United States.

Jones would take over the original Curtiss Bicycle Shop—now nicknamed the "Industrial Incubator"—in the village as the headquarters for his aeronautical business, known as the Jones Airship Company, formerly the International Aero Vehicle Company, the following year.[14] The Curtiss Company now found itself handling the production of airship components alongside the cycles and engines. To facilitate this, Curtiss began buying or renting buildings around the village to compensate for the crowded conditions the company continually found itself in.

Jones, a professional artist and engraver by trade, caught the aviation bug in 1905 while living in Dayton, Ohio. There he designed and built an ornithopter, a huge flying machine that flapped its wings like a bird in order to fly. The machine lacked a sufficient engine to power it, so he shipped it to Hammondsport in September 1906 after contracting with Glenn to build the V-8. Jones soon followed, relocating his family to the village.

No sooner had Jones placed the order when the news was out to the public that the V-8 engine was underway and generating great interest. Glenn would display one at the New York Auto Show in early December.[15] A second engine would be built the same time as Jones's for an airship named "Columbia" owned by Capt. William Mattery of Chicago, Illinois. The engine was quickly produced, delivered, and installed on the Mattery airship in time to be exhibited at the Chicago Automobile Show on February 2, 1907. Five months later, Mattery would visit Glenn in Hammondsport, spending two weeks

Charles O. Jones

conferring with Curtiss and Baldwin about possibly relocating his aeronautical business to Hammondsport.[16] That venture never came about.

Since Jones had not paid for his engine at the time of placing the order, and to ensure the new V-8's power, Glenn made a radical decision to mount it on a modified motorcycle frame. Hoping to feed his appetite for speed and setting new records, Glenn would create a most unusual machine for the upcoming Ormond Beach speed trials in Florida.

"The Industrial Incubator." The Curtiss shop just before C.O. Jones moved in.

C. O. Jones and his unsuccessful ornithopter in front of the Curtiss dirigible hangar, Hammondsport, New York. Jones is in the craft, with Curtiss to the left.

THE GREAT V-8 AND THE ORMOND BEACH SPEED TRIALS

The V-8 engine as it appeared on a test stand

Clarence White, the Curtiss shop superintendent, was put in charge of assembling the air-cooled V-8 engines with Glenn's input. The engine for C. O. Jones, once completed, was run on a test stand before being installed on the cycle frame and not tested again. Henry Kleckler (a new employee and later a major asset for the company in engine design) was put in charge of the night shift crew building the cycle and was given the task of creating and installing the beveled gears and universal joint for its driveshaft. He was also responsible for overseeing Charlie Wixom, the company carpenter, in the shipping department and with getting the cycle ready for shipment to the speed trials at Ormond Beach (Glenn's second and last visit), being held the fourth week in January. The finished cycle was huge—7 feet long and heavy, weighing around 200 pounds. It was equipped with an automobile tire in the rear and a cycle tire on the front.

According to Kleckler, once the cycle was completed

> there was considerable speculation how it would handle on the road. It was not possible to road[-] test such a thing and also there was over a foot of snow on the ground. We cleared a path about ten feet wide up the steep grade between the shops and shoved the machine up to the top and Mr. Curtiss got on and we all ran and pushed it down the hill while he tried out the balance. It did not give much of a test. But he said he thought he could handle it where he had room and under motion.

The V-8 motorcycle was immediately crated and shipped at the last possible moment with two other cycles, accessories, and other items. Kleckler also noted that this was how they normally did things around the plant in "those days."[1]

Accompanied by Tank Waters and Thomas Baldwin, Glenn would run the "freak bike" (a nickname given to it and others, not built for commercial sale) along with other Curtiss motorcycles for various trials at Ormond. Since the V-8 was untested, it proved to have a number of problems, so Glenn left it to Waters to work out the bugs while racing the other cycles first.

According to Tank, he had to get the handle grip twist throttle to work correctly, since it controlled two Curtiss-brand carburetors located on either side of the engine. Originally designed for smaller engines, the Curtiss carburetors proved to be inefficient, not allowing proper gas flow. This setback was corrected by replacing them with two larger Schebler-brand carburetors that Baldwin managed to borrow at the event. Proper ignition also proved to be a problem. This was solved with the use of eight dry-cell batteries. The only way to attach them was to wrap them around the oil tank, under the seat, with friction tape. Unfortunately, in order for this to work the seat had to be removed. As a substitute, a luggage rack was attached over the rear tire for Glenn to sit on. After several long days of fine-tuning, the monster was finally ready to run.[2] It should also be noted that the cycle also had a handle grip twist spark advance.

Obviously not a standard cycle, the V-8 could not be entered into any race on the docket. However, officials did agree to a time trial covering 4 miles on the evening of the twenty-fourth. This was to be overseen by a team using stopwatches. The group leader in charge of the team was Mr. James T. Sullivan, an event official.[3] Allowing a 2-mile start to get up speed, Curtiss managed to get the engine started, and the cycle proceeded to lurch forward while Baldwin and Waters tried running alongside to help Glenn keep his balance.[4]

Tank Waters steadying the V-8 while Curtiss works on the front wheel before the time run

Curtiss prepared to make his run

As he gained speed, and to cut wind resistance, Glenn leaned forward until he was practically lying flat on the bike (this was a practice he originally learned while racing bicycles). He covered the next timed mile in 26.4 seconds—136.4 miles per hour! It took him another mile to bring the bike to a stop. Minutes later, he would learn that he had shattered the world speed record, becoming the fastest man on Earth—going a mile per hour faster than any man in history.

In order to make the time an official standing world record, Glenn was required to complete a second run as fast as or faster than the first. After refueling, he took off again. He managed to build speed up to 90 mph when suddenly the universal joint on the driveshaft broke, causing it to flail about, bending and damaging the frame of the cycle. Miraculously, he was able to stop without injury to himself.[5] While Glenn was disheartened and a bit shaken over the damage to his motorcycle and not completing the second run, most everyone there that day agreed that he had set the record. The press proclaimed that on that day, only a speeding bullet could rival him.

Glenn's personal comments on the record-breaking V-8 run were a bit subdued due to his typically modest nature. In a magazine interview, he was quoted as saying, "Riding an eight-cylinder motor cycle is not likely to become very popular," and at such a high speed, "all I could see was a streak of beach with wild surf on one side, sand hills on the other[,] and a black spot where the crowd was. The machine set up a terrific and inexplicable vibration; it was so great that it did not create wholly comforting thoughts." However, he did say of the experience: "It satisfied my speed craving."[6]

Glenn Curtiss on his V-Twin motorcycle just after setting a world speed record, January 23, 1907

Enlargement of photo taken at Ormond, showing the "clutch" circled in red

Glenn H. Curtiss and Tank Waters with stuffed alligators in Ormond Beach, Florida, January 1907. Tank holds a bottle of celebratory champagne.

He also set an officially timed record of one minute, two seconds for the mile run in the single-cylinder class on January 22. Competing in the mile race for two-cylinder motorcycles on the twenty-third, Glenn came in first and yet again set another world record time of 46.4 seconds.[7] Both were medal winners.

Looking to set a fourth record on the twenty-fifth, Glenn and Tank were scheduled to make a try for the mile run with two riders on a tandem. Having installed a tandem attachment to the two-cylinder cycle, the pair arrived at the starting line, waiting their turn, when they witnessed the wreck of Fred Marriott's Stanley Steam Racer "the Rocket." While not fatal, the incident put an end to the speed trials, robbing the Curtiss/Waters team of a chance at another record.[8]

Baldwin headed out to California immediately after the event, while Glenn and Tank lingered, taking time to have a souvenir photo taken of them posing with a stuffed alligator. The two booked passage on a ship heading back to New York and subsequently ended up with an unpleasant case of seasickness due to rough seas.

Glenn made the initial decision not to return to the Ormond Beach speed trials after 1907 for one basic reason. He felt that the motorcycle events were poorly covered or ignored by the news media, with too much emphasis on the automobile.[9] However, another underlying factor may have been the unforeseen problem with the eight-cylinder cycle. While people thought of Glenn as a daredevil—and to some, a "Hell Rider"—Glenn did not take unnecessary chances, and safety was foremost in his mind. He always studied the courses he would run ahead of time as a precautionary measure and also scrutinized those whom he would be competing against. It would be fair to say that the second run of the V-8 was more of a risk than he was willing to take.

Photo of V-8 taken around 1970, while the cycle was at the Smithsonian; note the round "clutch" attached to the engine sans the driveshaft and gear.

World's Record — Ormond Beach, Fla.
1 Mile — 26 2/5 Seconds.
8 Cylinder, 40 H.P. Motor Cycle,
Built by the Curtiss Manufacturing Company, Hammondsport, N. Y.

Souvenir postcard published in September 1907. The colorized artwork was done by Charles O. Jones.

On his return to Hammondsport, Glenn suspended all work for the night shift employees responsible for constructing the cycle and entertained them with his experiences while serving them all a treat of fresh oranges.[10] The V-8 engine was removed from the cycle and turned over to C. O. Jones for his experiments and replaced with a dummy engine. The cycle was also repaired and made to look as it did before it was sent to Ormond, in order to be ready to fill requests for display at various automobile shows.

Over the years, stories have arisen as to how the V-8 was actually started. While documented details were never recorded in period accounts that can currently be located, it has been speculated by early biographers and over the last few years by motorcycle enthusiasts (some even publicly doubting the run ever happened) that the cycle was either towed by a car or pushed to start. As stated before, Tank specifically said that he and Baldwin ran alongside, not pushing.

It has always been suggested that the cycle was direct drive without a clutch. If so, the rear wheel could have been elevated, either with a portable stand or by two men lifting the back end off the ground in order for the wheel to spin freely when the engine started. The combination stand / luggage rack—which can be seen on the back of the cycle in the beach run photographs taken during the event—was an item that Glenn offered in his retail catalog, and may possibly have been used as intended. However, contemporaneous photographs from 1907, right up to 1990 photos, also show a large, round object just behind the engine and bolted to the frame. This may actually have been a gearbox or a centrifugal clutch, similar to those introduced in 1889 for use on early Daimler automobiles.

There is an additional aspect of the V-8 motorcycle that has seemed to have gone overlooked through the years. That is the possibility that a second V-8 cycle was built strictly for exhibition. The *Hammondsport Herald* reported on November 20 that "at the works just now is also being completed a monster eight[-] cylinder motorcycle[,] practically a counterpart of that which Mr. Curtiss made a world's record at Ormond Beach." This motorcycle was to be displayed at the Chicago Auto Show being held during the first week of December. Harry Genung, Curtiss plant business manager, was the representative in charge at that weeklong event.[11] However, a V-8 motorcycle had already been displayed at the New York Auto Show the first week of November.[12] This article raises an interesting question: Are there actually two V-8 cycles?

Assuming there was only one V-8 cycle, it would sporadically go on tour over the next few years, starting with the New York Auto Show held November 2–9 and then the Chicago Auto Show in December.[13] It would appear again in 1909 at the Wanamaker Department Store in New York City, and then at the Dayton Dry Goods Store, in Minneapolis, Minnesota, in 1910, along with the 1909 "Rheims Racer" airplane at both locations.[14] The cycle was last featured at the third annual Binghamton Auto Show in Binghamton, New York, at the Charles Wakeman booth, and also at his dealership in Lestershire (now known as Johnson City, New York) in February 1912.[15]

Eventually, the V-8 was moved from Hammondsport to Buffalo sometime during the early years of World War I, and then over to Long Island in 1917. There it was put on display in the front lobby of the Garden City facility until 1920, after which it quietly disappeared until 1929, when it was photographed with a Curtiss-Chieftain radial engine for size comparison.

At some point, the V-8 returned to Buffalo, where, during World War II, it was discovered in the storage area of the Flexlume Plant of the Curtiss-Wright Corporation in January 1943. Unfortunately, the *Curtiss Wright-er*, a company newspaper, printed a not-so-factual story identifying the cycle as being constructed in 1901. They also stated that it had set a record of 108 mph on Long Island while racing against two French cyclists the following year. The article went on to report that Glenn personally used it to commute from his home to the factory on Long Island until 1908.[16] Sadly, this shows that even a company can get its own history wrong.

After its reemergence from storage, the motorcycle would be brought out occasionally to display at special company events. Curtiss-Wright employee George A. Page Jr., chief engineer, saw to it that the cycle was donated to the Smithsonian Institution in 1952,[17] where it currently resides. It should also be noted that the Schebler carburetor that appears on the cycle in later years is a Model H, not available until around 1914.

The original V-8 did, however, return home to Hammondsport, New York, in 1975 for exhibition at the Glenn H. Curtiss Museum, where it remained until 1988. It was reloaned from 1992 to 1995, so that an exact replica could be built by the Mercury Aircraft

The V-8 motorcycle posed with a Curtiss Chieftain Radial Engine in 1929 at the Garden City facility

Company of Hammondsport and museum volunteers for permanent display at the Curtiss Museum. At some time during its return to the Smithsonian between 1988 and 1992, the "transmission" was removed and the universal joint, shaft, and gear were added.

In 1996 the Smithsonian conducted an examination and conservation of the cycle in their possession, with these initial findings offered to the Curtiss Museum:

(1) X-ray and chemical analysis showed only the frame was original,

The original V-8 motorcycle while on exhibit at the Glenn H. Curtiss Museum in 1992, now missing the clutch and with an added driveshaft

(2) the gas tank was a replacement, and

(3) the engine was a shell with used defective cylinder heads.

In addition, when several layers of paint and reproduction decal were removed, the original Curtiss decal on the gas tank was revealed, as well as a "red racing decal" (no description) on the front column.[18]

Also changed during the Smithsonian's conservation, the black-tubing handlebar was replaced with a nickel-plated one, and the missing twist grips were added. In addition, the vintage black tires on the machine were switched to white tires to represent those originally used at Ormond.

In 2021, the cycle would undergo a second conservation at NASM before being installed in a display titled *Nation of Speed*.

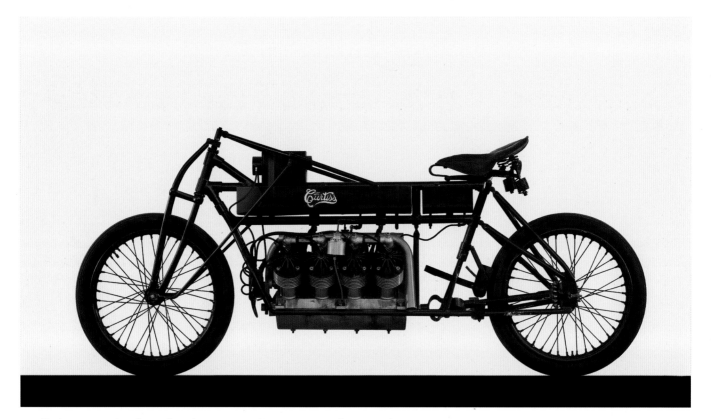

Replica V-8 motorcycle at the Glenn H. Curtiss Museum. *Photo: Mark Preston*

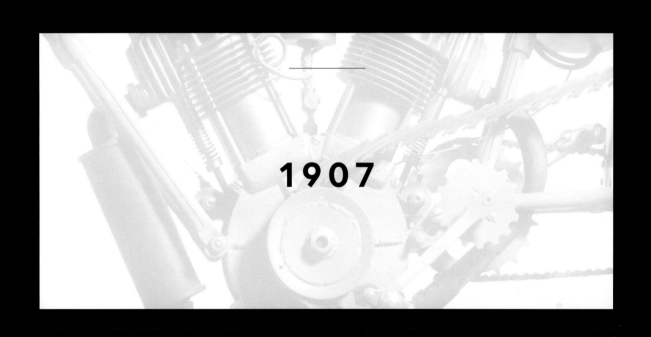

CHAPTER 10

1907

Interior of the Curtiss shop in 1907, with Glenn standing in the back with a new four-cylinder, in-line aeronautical engine

Glenn began taking more of an active role as an engineman at Baldwin's dirigible events. This resulted in Glenn personally building the first privately owned airship hangar in the United States that May, to house Baldwin's dirigibles.[1] This was located on property leased by Glenn known as Kingsley Flats, a large, flat plain east of Hammondsport at the head of Keuka Lake.

His next race for the year would be at the Memorial Day event sponsored by the combined efforts of the New York and Brooklyn Motorcycle Clubs, located at Manhasset Hills, Long Island, on May 30. Taking both a single- and two-cylinder machine with Albert Cook as a companion racer, the two would walk away with all three prizes offered at the event. Cook would take first on the single, while Glenn took first place in the Class III event and first in the Free-for-All class on his V-Twin.[2]

Still fascinated with cars, Glenn took on the Ford agency and promptly sold L. D. and Victor Masson a six-cylinder, 10hp touring car, after which he purchased one for himself a few months later.[3]

Tank Waters offered his Erie in a variety of styles as a complete motorcycle or kit featuring 2 or 3hp, one-cylinder engines in friction or belt drive. In December, a one-cylinder, 2hp, two-stroke Curtiss model A1 engine with Curtiss carburetor specifically designed for bicycle conversions was added.[4] However, Tank also obtained engines from the Reliance Motor Cycle Company of Addison, New York. According to Tank, only the 3hp engines had the name "Reliance" on them. This caused confusion at times regarding who produced the 2hp engines, since they resembled each other and all were advertised as Eries.[5] At this time, Tank introduced the Erie brand of spark plug.

Having received his V-8 engine from Curtiss upon return from Ormond Beach, C. O. Jones installed it in his first of three attempts at flying. The contraption was known as an ornithopter, which flapped its wings in order to fly. Needless to say, the experiment failed, since the engine was too powerful for the aircraft, and it literally beat itself to pieces during tests.

Glenn would enter into a partnership with Baldwin this year in order to sell a dirigible to the US Army. On June 27, Baldwin gave a demonstration to the army and a crowd of several thousand people in Hammondsport at Kingsley Flats during Gala Day (billed as an early Fourth of July celebration) in his new dirigible, the "Twentieth Century" (the first to be constructed at the Curtiss plant). The dirigible was powered with the new Curtiss 20hp, four-cylinder, air-cooled, in-line engine, strictly designed for aeronautical use, propelled by counterrotating propellers. Baldwin flew out over Keuka Lake, thrilling the crowds.[6] No longer able to resist the temptation, Glenn had to try flying the dirigible for himself. He took the challenge the following day after promising Baldwin he would not try any stunts.[7] Able to control the machine smoothly without incident, Glenn stayed aloft for twenty minutes. His response when asked what it was like to be in the air was "It is delightful, only there is no place to go." Moreover, "There is absolutely no fear of falling."[8]

One local photographer who was there to record the event was none other than a close friend of Glenn's, Harry Benner. Harry had taken up photography as a hobby around 1888 and became a professional photographer, opening a studio in Hammondsport in 1906. He immediately began recording all the motorcycle and aeronautical experiments taking place in the area. Glenn himself had in the past tried to photographically record most of his experiments as well as his various motorcycles when possible. With Benner now on the scene, Glenn made him the official photographer for the company.

Dirigible hangar on Kingsley Flats on Lake Keuka, with Glenn Curtiss standing in front

Glenn H. Curtiss in his first dirigible ride, piloting the "Twentieth Century," June 28, 1907

Another view of Curtiss in the "Twentieth Century," June 28, 1907

Benner would produce a large number of postcards of the various experiments and offered them for sale to the public through local retail establishments. The process became so profitable that Glenn had his company representatives selling Benner photo postcards while traveling around the country. Benner had stated that just in one month during those early years, he had produced and sold 92,000 postcards alone. All this

would end with the United States' involvement in World War I. Harry would enlist, go through advanced photo school at Cornell University, and be commissioned a second lieutenant in the Signal Corps photographic division. Lieutenant Benner would remain in the army for three years after the armistice in 1918. After practicing photography in Ohio in the 1920s, he retired to Hammondsport in 1932, passing away in 1946.

Glenn H. Curtiss and Captain Thomas S. Baldwin

Harry M. Benner

The Curtiss baseball team posed with Curtiss motorcycles. The lineup includes both one- and two-cylinder cycles in black and gray.

During the Gala Day celebration, Curtiss sponsored a 10-mile motorcycle race as one of the many events coinciding with Baldwin's flight. While it was not reported how many competed, the top prize of a solid-gold medal was won by Charles Kirkham riding his own designed motorcycle, beating out the front-runner, Albert D. Cook, the latter having fallen due to the muddy road conditions caused by rains the night before.[9] Sylvester Carroll came in second; Will Damoth, third; and David Brandow, fourth. While not placing, Cook received the Time trophy. A motorcycle hill climb that was scheduled during the gala had to be canceled due to muddy and hazardous conditions.

Meanwhile, the employees were busy organizing their own baseball team, known as the "G.H.C.s," with Frank Taylor as manager. Posing in their new uniforms after a hard-won game, the team would have a group photo taken, each with a Curtiss motorcycle. The photo was titled "Winners on winners." A similar photo was featured on the cover of the January 1908 issue of

The Curtiss baseball team as pictured on the cover of the January 1908 issue of *Motorcycle Illustrated*. A. D. Cook stands with a Curtiss cycle.

Motorcycle Illustrated magazine. The team challenged many of the other business teams throughout the county, such as the Hoyt Cooper Shop and the Urbana Wine Cellars, and the competition was well supported by the community. The team consisted of Jay Douglas (pitcher), Will Damoth (catcher), H. Horton, Jay Safford, Thomas Hunt, Wheeler Hunt, Frank Taylor, Dee James Davis, C. Willard, Lucius Lowry, Albert Cook, John Osborne, and Harry Longwell, as well as Anna Swarthout, company stenographer, as mascot.[10]

A few days after his first ride in an airship, Glenn was in Clarinda, Iowa, assisting Baldwin on July 4. Baldwin set a long-distance flight record at the event.[11] In Glenn's absence over the Fourth of July, five Curtiss employees, including A. D. Cook, Tim Losey, Frank Taylor, D. Brandow, and S. Carroll, represented the company at the Independence Day motorcycle race held in Penn Yan, New York. Cook would take first-place awards in both of the 3-mile races, with Losey and Taylor each taking a second place.[12]

Glenn had found over the last couple of years that the business, the shows, aiding Baldwin, and racing were taking him away from home more than he liked. His schedule for the month of July alone would find him in Iowa, Nova Scotia, Rhode Island, and New York City, with a stop or two back in Hammondsport. August was just as busy. To help ease the pressure, Lena gave up managing the business office, the job being turned over to longtime friend Harry C. Genung. This allowed her to increase her time traveling with Glenn, to ensure that he would take care of himself and keep their relationship going strong. Lena would, however, stay on part time as office help until April 1910.

Glenn's last motorcycle competition for the year took place in Providence, Rhode Island, during the Annual Federation of American Motorcyclists meet, which ran from July 31 to August 3. He would take a total of three first- and one second-place trophies, all

the while setting another world speed record at the event. These included first place both in the Free-for-All Hill Climb and One Mile Time, a world record of 56.4 seconds, for the "30.5 Cubic Inch" class (one cylinder), as well as first place in the One Mile Time and second in the Free-for-All Hill Climb for the "61 Cubic Inch" class. This would prove to be his best haul of trophies at any one event during his racing career. Albert Cook, his racing partner, would take "Over All First" in the endurance run from New York to Providence on a two-cylinder cycle, winning a gold and diamond medal for the two-day-long race, as well as second place behind Glenn on the Hill Climb for the "30.5 Cubic Inch" class. The Curtiss team would prove hard to beat.

These laurels, however, came at a cost. Glenn would start the event off with a scare. During one contest the carburetor began leaking, resulting in the gasoline catching fire and threatening serious damage to himself and the machine.[13] And at the end of his last race, the one-cylinder Free-for-All Hill Climb, Glenn was injured while overshooting the finish line, resulting in hospitalization.[14] A knee injury would require the use of a cane for several months, and a painful hand laceration made it difficult to take an active part in shopwork. As a result, Glenn forbade Cook from continuing with the rest of the event, finding the course too dangerous. With two near misses and now a major injury, all within six months, Glenn agreed to stop racing at the request of his wife, Lena, who had been present at the event.[15] With his racing days presumably done, Glenn allowed others to represent the company in his place.

As to the Curtiss motorcycle staying in the forefront of the racing circuit, several other Hammondsport natives and company employees would continue racing for Curtiss aside from Albert D. Cook. Will Damoth, also a childhood friend, employee, and early bicycle racing partner of Glenn's, would see to it that the legacy continued, races were won, and records were set.

Gold-and-diamond medal won by Albert D. Cook for the FAM Endurance Run at Providence, Rhode Island

Glenn just after setting a world record of 56.4 seconds for a mile at Providence, Rhode Island, July 1907

Glenn holding a four-cylinder, in-line aeronautical engine; taken in July 1907

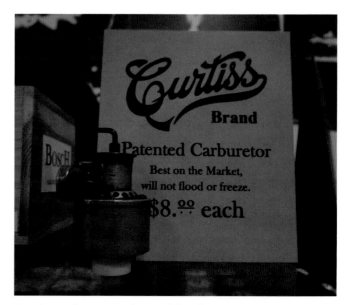

Curtiss carburetor. *Photo: Richard Leisenring, Jr.*

Independent racers around the country were also enthusiastically using Curtiss cycles by this time, ensuring that the Curtiss racing heritage continued. Some independents that would win trophies riding a Curtiss cycle included A. Carlton, F. E. Carlslake and George A. Faulkner.

Another such supporter was Henry J. "Hank" Wehman. Wehman, an avid racer and the secretary of the Federation of American Motorcyclists, would become a Curtiss distributer in New York City at the end of 1908 under the name of the Curtiss Motorcycle Company (no relation to the G. H. Curtiss Manufacturing Company). Hank, like Glenn, was a tinkerer and designer. His first creation of notoriety was a tricar in 1906. The machine was constructed using a Curtiss V-Twin motorcycle with an Indian fore car attached to the front in order to put the passenger in the front of the vehicle.[16] Wehman later worked closely with Glenn, designing a new style of frame known as the "Wehman," which the G.H. Curtiss Manufacturing Company would offer starting in 1909.[17] By 1911, Wehman would close his dealership, move to

CARBURETOR

Price, $8

Capacity. For engines having dimensions not exceeding 3½x3¼", ¾" pipe connection

This carburetor is central draft with a central gasolene supply. Perfectly automatic. Has an adjustable automatic air supply, and is better adapted for motorcycle use than any other carburetor

It will not shake up and give bad mixture while going over rough places; will not flood, slop gasolene, and will not freeze in cold weather

This carburetor is not only very neat in appearance, but is light and compact. It weighs but 15 ounces, and embodies all the most up-to-date features of scientific carburetor construction

Advertisement for the Curtiss carburetor, 1907

Curtiss Buckboard prototype, 1907

Hammondsport, and work for Glenn, becoming one of the chief office men in the Curtiss plant both for motorcycles and later aircraft.

Engine production was at an all-time high at this time, Curtiss having received a contract from the War Department for V-Twin engines as generator motors. Municipalities such as Rochester, New York, were looking into or purchasing Curtiss motorcycles for their police departments.[18] Now employing forty-six shopmen, exclusive of the business office,[19] the company had orders for seventy-five machines due by April 1 and another four hundred machines being prepared.[20] It is also estimated the company produced between five hundred and six hundred motorcycle and aeronautical engines of various types that year. The main plant now covered over 25,000 square feet and was producing eight complete motorcycles a day.[21] The growth of the company was so extensive that it caused a major housing shortage in the village.[22] An interesting sidenote to this growth is that the *1907 Curtiss Motors & Motorcycles* catalog stated for the first time that the company also manufactured their own leather belts and tool bags—an indication that they now had a leather shop as part of the plant, thus creating a need for other skilled craftsmen not previously used.

The four-cylinder in-line and the V-8 engines were quickly added to the catalog and in immediate demand. Other additions to the catalog were his own brand of motor oil, as well as the Curtiss-designed carburetor (introduced in 1906), which would also be a special feature on all Curtiss cycles for the next few years.

Having held the agency for the Orient Buckboard for a number of years and making several sales, Glenn toyed with the idea of making his own version of this automobile utilizing a V-Twin engine. While a Curtiss Motor Car Co. was formed,[23] the Curtiss Buckboard, which appeared only as a prototype, was never put into production. Photos of it were taken to add to the 1908 catalog but were pulled before it went to press.[24] The company would also experiment with various other vehicles such as a railroad motorcycle.[25]

Assisting Baldwin first in New York City in mid-August with airship demonstrations, Glenn would again accompany him to Halifax, Nova Scotia, Canada, in September for several more dirigible demonstrations.

A slight diversion would take place for Glenn during the month of October while back in Hammondsport. Charles O. Jones, having just completed his second flying machine, would launch it on the twelfth as a glider off the ridge to the east of the village, with Glenn's assistance. This design was unique since the wings folded into the fuselage like an accordion. The flight of this unusual aircraft was believed to be the first heavier-than-air flight in Steuben County and certainly a taste of things to come.

However, seeing the success of Baldwin and his dirigibles caused Jones to change his focus, this time utilizing the Curtiss V-8 on a 95-foot-long dirigible named the "Boomerang" *that* he would construct with the help of Glenn and Thomas Baldwin later in the winter of 1907–08.

Glenn was off again a few days later to assist Baldwin at the St. Louis Air Meet in Missouri to compete in the dirigible races there. Baldwin took third place in his dirigible the "California Arrow," having being beaten by his former employee Lincoln Beachey.[26] Since Baldwin was using a Curtiss four-cylinder, in-line engine, Glenn acted as the mechanic and crew manager for the event.

After a short rest back in Hammondsport, Glenn was then off to the New York Auto Show the first week in November to display the V-8 motorcycle and other machines. Accompanying him was his wife and Tank Waters, with Harry Genung as an assistant. Harry would handle most of the business at the show, giving Glenn and Lena a chance to enjoy a short, most-needed vacation together. [27]

Shortly after Glenn arrived home from New York City, Harry C. Gammeter of Cleveland, Ohio, sought him out for advice on how to solve the problem of powering an ornithopter he had designed and built the previous year. While Glenn was somewhat skeptical of the concept of the ornithopter after the Jones failure, he agreed to help, and the machine was immediately shipped to Hammondsport, where a temporary building was erected next to the Curtiss plant to house it for further indoor experiments during the winter months.

After some consultation over how to power the machine, Glenn and Gammeter settled on trying a Curtiss 7hp V-Twin aeronautical engine due to the light construction of the craft. It was feared that a more powerful engine would tear the aircraft apart. After several attempts, the ornithopter successfully, although briefly, lifted off the ground in an indoor, unmanned, tethered flight. While satisfied with the outcome, Gammeter put aside the experiments due to poor health and returned to Ohio. The machine would remain in storage at the Curtiss plant as late as March 1908. In December 1909, Harry Gammeter announced his intentions of pursuing additional experiments with the machine but never followed through. While he initially filed for a patent in October 1907 with no mention of a Curtiss engine by name, the final document granted on August 30, 1910, clearly shows a Curtiss-type V-Twin engine in the drawings.

Harry Gammeter would go on to develop Green Springs, Ohio, as a health spa and, while never having the opportunity to revive his experiments or fly, was a member of the Aero Club of America and served as president of the Aero Club of Cleveland for several years. Gammeter would pass away in 1937 at the age of sixty-seven.

On November 26 in Hammondsport, Glenn would intentionally set his very first aeronautical record while demonstrating the durability of his engines to the US Army. Aware of the standing endurance record, Glenn felt it necessary to break it to help secure the army dirigible contract he and Baldwin were working on. Glenn remained aloft in a dirigible for four hours under continuous motor power while gaining speeds of 30 mph and an altitude of several hundred feet. During his flight, he periodically dipped low enough to the ground to talk to spectators. As a result, Glenn broke the previous endurance record by one hour.[28] In the meantime, Tank Waters was in the process of arranging to have a small building he had just purchased relocated to the Curtiss property for a storehouse.[29]

A major directional shift for Glenn and his company began to take place this year as a result of his interactions with Dr. Alexander Graham Bell. Glenn had received a second engine order from Bell at the December 1906 New York Auto Show. The engine that Bell purchased was for flying-machine experiments, which Glenn would personally deliver later in July.[30] In January, Glenn would briefly visit Bell in Washington, DC, on his way to Ormond Beach. As 1907 came to a close, Glenn sold the prized V-Twin used at Ormond Beach to Waters, who would use it for several years before disposing of it.[31]

Charles O. Jones's heavier-than-air machine that flew as a glider on October 12, 1907, in Hammondsport, New York. Jones is seated in the machine, with Curtiss off to the far right.

C. O. Jones standing inside the gasbag of the dirigible "Boomerang"

The Gammeter ornithopter

CHAPTER 11

THE AERIAL EXPERIMENT ASSOCIATION

Dr. Bell enticed Glenn to join his newly formed Aerial Experiment Association (AEA) late in September 1907, after Glenn made his visits to Bell's home in Bedeck, Nova Scotia, Canada. Reported Glenn, "The object of the organization was to fly a successful flying machine and get into the air."[1]

It is not the intention of this book to go into full detail about the doings of the AEA; this chapter is only to show the amount of time Glenn Curtiss invested in the group, pointing out the direction that he and his company were destined to head in.

Glenn reluctantly joined at first. While still running the G. H. Curtiss Manufacturing Company, working with Thomas Baldwin on the US Army dirigible contract, and helping C. O. Jones with his experiments, Glenn was now fully involved with the AEA and the development of heavier-than-air flight. Originally formed in Bedeck, the association moved to Hammondsport later that winter to take advantage of the Curtiss manufacturing facilities.

The association would consist of Bell and four men—J. A. D. McCurdy and Frederick W. "Casey" Baldwin (no relation to Captain Thomas Baldwin); two young Canadian engineers; Lieutenant Thomas E. Selfridge, an official representative of the US Army; and Glenn. Glenn would be appointed director of experiments and engine expert.

After some initial experimentation with a glider, each man was given the opportunity to design, construct, and test—with the group's input—a flying machine. These machines, known as aerodromes, fell into this order:

(1) the "Red Wing," Thomas Selfridge, tested March 17, 1908

(2) the "White Wing," F. W. Baldwin, tested May 18–23, 1908

(3) the "June Bug," G. H. Curtiss, tested June 21–July 10, 1908

(4) the "Silver Dart," J. A. D. McCurdy, tested December 6, 1908–February 24, 1909

When the group first met in Hammondsport to proceed with their aeronautical experiments, Glenn presented each of the men with a motorcycle in order to travel around the village and local area. As expected, as soon as the weather permitted, the men could be found periodically racing each other on the cycles when not involved in aerial experiments. For long-distance errands and Dr. Bell's convenience when visiting Hammondsport, the association purchased a 1908 Peerless four-cylinder automobile through Curtiss.[2]

To show the fascination "Bell's boys" had with, and the durability of, the Curtiss cycles, Charles A. Champlin liked to tell an anecdote of "Casey" Baldwin and J. A. D. McCurdy that took place in June 1908. The two were out on a joyride on a tandem and dropped by to say hello. Noticing that McCurdy was wearing only bedroom slippers, the two said they were out for a quick spin, since it was a lovely day. Later that night, Champlin received a long-distance call from the manager of the Hotel Buffalo in Buffalo, New York, asking if he would vouch for two scruffy young men on a motorcycle. Shocked, Champlin spoke to Baldwin and asked if they had taken the train, since the distance was over 100 miles. Casey replied, "No, the roads were so good and the cycle was running like a watch[,] so we just kept going." He asked Champlin to please "let Glenn know where we are."[3]

The AEA—G. H. Curtiss, F. W. Baldwin, Dr. A. G. Bell, Lt. T. E. Selfridge Jr., and J. A. D. McCurdy

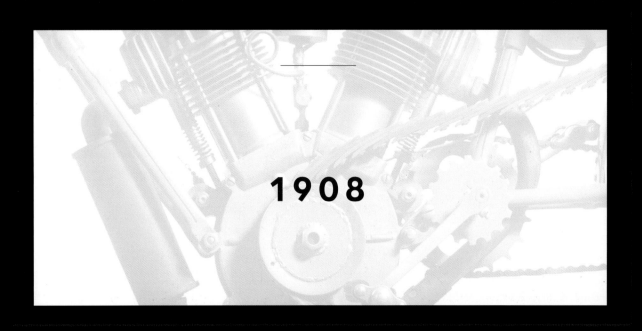

CHAPTER 12

1908

One customer, Professor J. Newton Williams, had purchased a 20hp, four-cylinder, air-cooled, in-line engine from the Curtiss Company a few months prior for a helicopter experiment in Ansonia, Connecticut. The engine unfortunately proved to be insufficient, and after contacting the Curtiss Company, Williams received a personal invitation from Glenn Curtiss to bring the machine to Hammondsport, where he could use a 40hp, eight-cylinder, air-cooled engine free of charge.

Arriving in January 1908 with his machine, Williams began assembling the helicopter in the dirigible hangar that had been erected a few months before near the lake on the outskirts of the village. This was a very opportune time, since Dr. Alexander Graham Bell was here with the new Aerial Experiment Association, from which Williams received much encouragement. After assembly, adjustments, and several trial runs, the Williams helicopter was brought out for its official testing on May 22. Tethered to the ground, since it had no steering system, the machine succeeded in lifting a passenger, a young Curtiss employee by the name of Byron Brown, off the ground for a height of 40 inches to become the first vertical liftoff and flight of a manned flying machine in the United States.

Williams helicopter

C. O. Jones's Boomerang in the glen at Hammondsport, New York

On June 16, 1908, C. O. Jones would take his new Hammondsport-built dirigible the "Boomerang," with the Ormond Beach–used V-8 engine, out on its maiden flight. The "Boomerang" was the third of three attempts at building various flying machines for Jones. The airship had been readied on the west side of the village, at the mouth of the Hammondsport Glen next to the old Mallory Mill. The "Boomerang" was unique in that it was designed to carry passengers and tow a smaller baggage balloon. One other innovation that was a departure from other dirigibles was a sliding ballast system, allowing the operator to remain seated in the undercarriage or gondola while getting the airship to rise or descend without having to move back and forth by using his own weight to operate the ship.

For the event, Glenn was merely a bystander, Jones was in charge with Albert D. Cook assisting. As the dirigible lifted out of the Hammondsport Glen, the steering gear broke, allowing the ship to drift toward and strike the bank, which caused the gasbag to sag. Jones shut down the engine as a safety precaution. Rising slowly over the Curtiss Company plant, the "Boomerang" headed southeast, sailing smoothly for thirty minutes until Jones brought it down near the village of Savona. While the gasbag was torn on landing, the flight was considered a success, with the engine proving its worth.[1]

Jones moved the airship to Binghamton, New York, for an exhibition tour and then headed to Palisades, New Jersey. After several successful demonstrations, the "Boomerang" began to be plagued with small mishaps, which caused several minor crashes.

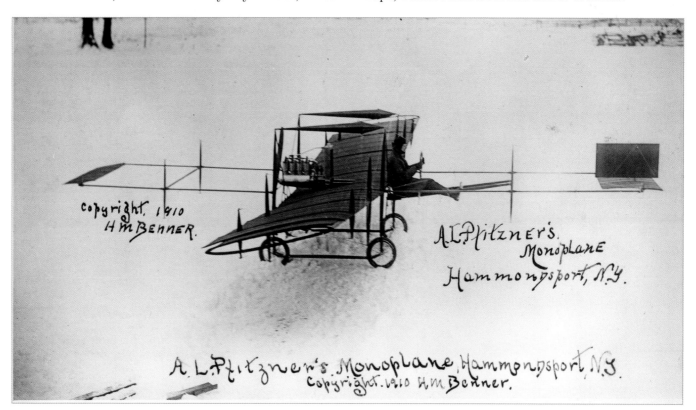

Pfitzner monoplane with Alexander Pfitzner seated at controls

Heading next to Waterville, Maine, Jones made a couple of successful practice flights before his main exhibition for a crowd of 25,000 on September 2. On that day, Jones had a problem with getting the gasbag to properly inflate, causing the dirigible to sag. Deciding to proceed anyway, he took the airship up, finding the sag worsening as he gained altitude. The underinflated bag was sucked into the propeller, ripping it, with the heat from the engine igniting the gas. The resulting explosion hurtled Jones to his death.

A V-Twin aeronautical engine set up in a dirigible

Sadly, because the family was deep in debt due to his dreams of flight, Jones's remains could not be brought home and were buried in a pauper's grave in Waterville. The wreckage of the "Boomerang," tent, and equipment was sold to pay outstanding expenses.[2] His widow and three children returned to Cincinnati, Ohio. The incident was rumored to have a profound effect on Glenn, having worked closely with Jones. His penchant for safety increased significantly.

To this day, the fate of the historic V-8 engine is shrouded in mystery, and what became of it was never publicly mentioned.

Earlier in the year, a young Hungarian army officer by the name of Alexander L. Pfitzner came to work for Curtiss as an engine designer. Having heard of his talent and ability to improve engines, Glenn had managed to hire Alexander away from the Buick Automotive Company. Building off his work at Buick, Pfitzner developed a water-cooling jacket system for cylinders later in the year. As a result, the air-cooled engines for aeronautical applications quickly became obsolete.

Simply explained, water or liquid is circulated around the cylinders of the engine while running, to keep the engine cooled regardless of whether it is in

June Bug in flight, July 4, 1908

Scientific American trophy

movement or not. This reduces the heat that would cause expansion of metal parts, creating friction. As a result, this improves efficiency. Air-cooled engines require air to be forced around the cylinders while the engine is in motion to achieve the same result, but efficiency is reduced when the engine is running while not in motion. This allows the engine to overheat and possibly seize up.

One other innovation that Pfitzner applied to Curtiss aeronautical and later motorcycle engines was an overhead valve system.

Within a matter of weeks, Pfitzner developed such a keen interest in flight designs that he came up with a radical new method of lateral control, which utilized sliding wing panels to turn the aircraft. Wanting to build the plane of his design, Pfitzner struck a deal with Glenn to have it built in Hammondsport as a part of his wages with the company. While the plane followed the AEA patterns in many respects, the main differences were that it was a monoplane and utilized the sliding panels instead of ailerons to control lateral flight.

By February 1910, the monoplane was built, and initial flight tests were made on the frozen surface of Keuka Lake. Since Pfitzner had never learned to fly, his first flight resulted in some damage to the machine, but he gained experience in handling it with each successive attempt. By mid-April 1910, Alexander had made over twenty flights in the monoplane. While a few ended with serious damage to the plane, he always managed to walk away unhurt.

For personal reasons, A. L. Pfitzner cut ties with Hammondsport and Curtiss in late May to work with the Burgess Company in Marblehead, Massachusetts, taking the monoplane with him. He applied his sliding panel control to their biplane with only partial success. After completely demolishing the Burgess plane on July 8, barely escaping with his life, Pfitzner became despondent. After destroying all of his papers pertaining to his aeronautical devices, he took his own life on July 12, 1910. Oddly, it was left to Glenn Curtiss to notify Pfitzner's family in Hungary of the tragedy.

The one-cylinder engines could now be found in the 1908 catalog not only as an "Airship Motor" in 3 hp, but as a 2hp, two-cycle version for bicycle conversions known as the "A1." A 3hp, four-cycle version for motorcycles was also offered.[3] It appears that this is the only year this engine was offered as two- and four-cycle versions, since they are not listed in later catalogs or advertising.

Glenn's personal involvement with dirigibles would come to an end this year. Having won the US Army contract earlier in the year, the project would culminate with the delivery and demonstration of the army's first airship, the "SC-1," the largest airship of its kind in the United States at the time. The newly designed Curtiss 30hp, four-cylinder, water-cooled, overhead-valve, in-line engine would power the airship. An interesting aspect of the actual engine used on the "SC-1" was that it was still considered a prototype. The block was an air-cooled four with the copper-jacketed cylinder units. The complete valve mechanisms were purchased from the Franklin Automobile Company of Syracuse, New York, in order to expedite production.[4]

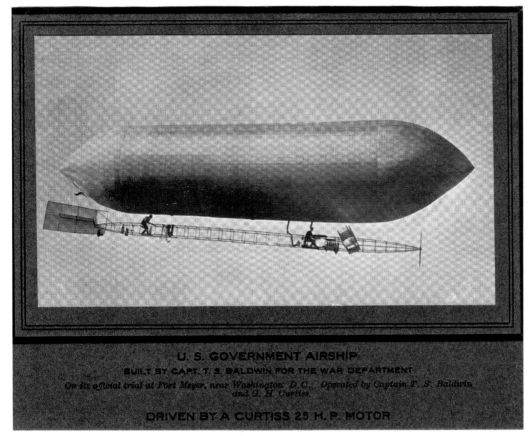

U. S. GOVERNMENT AIRSHIP
BUILT BY CAPT. T. S. BALDWIN FOR THE WAR DEPARTMENT
On its official trial at Fort Meyer, near Washington, D. C., Operated by Captain T. S. Baldwin and G. H. Curtiss
DRIVEN BY A CURTISS 25 H. P. MOTOR

Souvenir photo of the SC-1 on its maiden flight

On June 6, Glenn went against Lena's earlier wishes and participated in one more motorcycle race. Entering the New York Motorcycle Club–sponsored Fort George Hill Climb in New York City and accompanied by Thomas Baldwin as well as Lieutenant Thomas E. Selfridge and J. A. D. McCurdy (both members of Bell's Aerial Experiment Association), Glenn competed in three of five events. He took second place in the Free-for-All on a V-Twin, third on a single-cylinder in the Class B event, and fourth in the Class D event on his V-Twin.[5]

A major event that would solidify the change in the direction of the G. H. Curtiss Manufacturing Company away from the production of motorcycles took place on July 4 in Hammondsport.

A. D. Cook just after his win on the endurance run from the Catskills to New York City in June 1908

The FAM Endurance medal won by Albert Cook in June 1908

In September 1907, *Scientific American* magazine had put up a huge 31-inch-tall trophy made of 218 ounces of sterling silver with an onyx base for various accomplishments in heavier-than-air flight. Turned over to the Aero Club of America, the trophy was to be presented annually by them for various required feats. The first competitor to win it three times in separate years would keep it permanently. The first competition requirement was to fly a heavier-than-aircraft publicly for a distance of 1 kilometer in a straight line.[6]

Having been successful with previous tests of the "June Bug," Glenn publicly announced that he would try for the trophy on July 4, with Bell's permission. That day, Glenn successfully flew a distance of 5,090 feet, exceeding the required distance. It was the first preannounced, officially recognized, and publicly observed flight in America. Glenn would go on to win the trophy two more times to take permanent possession of it. This event would overshadow the attempted run of the 1904 V-4 motorcycle at Kenilworth Park in Buffalo, New York (see chapter 7).

On August 4 at Fort Meyers, Virginia, assembly of the Hammondsport-built dirigible began with the help of three army officers—Lieutenants Thomas E. Selfridge, Benjamin D. Foulois, and Frank P. Lahm. An interesting note is that Selfridge, a member of Bell's AEA, designed the propellers for the machine. A few days later, Baldwin as pilot and Glenn as engine operator ran several test flights to ensure that the "SC-1" was ready for her final inspection. On August 15, the "SC-1" passed all the required endurance tests and was officially turned over to the Army Signal Corps on the eighteenth.[7] For Baldwin; these demonstrations would bring about a purchase from the German government for a similar airship powered by a Curtiss two-cylinder, 7hp engine. It was touted as the first airship to be exported from America.[8]

With the help of Glenn and company, Baldwin would build over twelve dirigibles in Hammondsport between 1907 and 1909, making the village a major airship production center. Eventually, seeing dirigibles as fast becoming outdated, Baldwin—with Glenn's help—would move on to experimentation with and construction of his own designed airplanes known as the "Red Devil."

Glenn would head to Buffalo to attend the founding meeting of the Motorcycle Manufacturers' Association on September 15, where he would be appointed temporary treasurer until such time as the charter and bylaws of the new organization were drawn up. Two of the goals of the organization were to ensure fair representation of the industry in government legislation and advance relations among manufacturers, agents, and riders.[9]

Albert Cook's victories as a representative of the Curtiss Company over the years were impressive enough to feature his awards in company advertising. The medal he won for the 1908 FAM Endurance Run from Catskill to New York City on June 28 and 29 was featured as part of the major advertising campaign seen in the trade journals that year. While placing fifth, he completed the event with a perfect overall score, earning him the award. Also at the event were two other Hammondsport boys. While not winning any awards, Will Damoth would place eleventh on a Curtiss, and C. P. Rudd, an employee of the Motorcycle Equipment Company, placed twenty-third on an Erie out of a total of sixty-five contestants.[10]

Cook, one of Curtiss's first employees, moved to Rochester, New York, to open a Curtiss distributorship on Monroe Avenue in late 1908 and began racing as a private individual. Among his accomplishments was the design of a cycle frame that would bear his name. The cycle would be publicly offered by Curtiss in 1911.[11] Cook eventually closed the Rochester shop in July 1910 with the demise of his supplier, the Herring-Curtiss Company.[12] Shortly after, Albert would help form the Flower City Motorcycle Club in Rochester in 1912, then move to Buffalo, New York, start a new motorcycle business, and continue racing in Buffalo-area events until 1916.

On October 24, at the Vanderbilt Cup Race on Long Island, Glenn served as a volunteer motorcycle courier.[13] Later, in November, he had made arrangements to attend the FAM meet in New York City, where the New York Motorcycle Club had decided to honor him with a formal dinner. However, due to an unspecified illness in the family, he had to cancel. The dinner nevertheless was held in his honor despite his absence and, as the papers reported, went off without a hitch.[14] Glenn's participation in motorcycle events was now greatly overshadowed by his growing involvement in aviation.

1909

1909 Curtiss motorcycle. Wehman frame with V-Twin 7hp engine. *Photo: Derek Snyder*

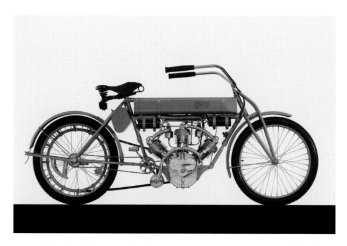

1909 Curtiss three-cylinder motorcycle. *Courtesy of Dale Stoner, photo by Mark Preston*

Even with Glenn now heavily involved in aviation, the Curtiss motorcycle was still extremely popular with riders and agencies across the United States. *Motorcycle Illustrated* magazine would call Hammondsport the "motorcycle hive" of the country, with all the related activity taking place in the tiny village.[1] While concentrating primarily on aeronautical engines, the company, which was now employing one hundred men, was offering five models of cycles in two frame styles, the Standard and the new Wehman.

The Wehman would be promoted as more of a high-end cycle and would sport a more powerful one-cylinder engine at 3.5 hp, as opposed to a 3 hp for the Standard and a 7hp rather than a 5hp two-cylinder engine.

The new, short-lived, 10hp, three-cylinder motorcycle known as the "Triplet" made its debut at the January 1909 New York Auto Show.[2] Unfortunately the cycle had issues, which kept it from performing any better than the two-cylinder model, and giving it bad reviews, which hindered sales. Only a few were ever made. Around three hundred of all models were estimated to have been sold by the end of the year.[3]

Aeronautical engines now ruled the G. H. Curtiss Manufacturing Company—so much so that they were also the engine of choice for aerial experimentation. One example was for a counterrotating, propeller-driven helicopter built by William A. Purvis and Charles A. Wilson of Goodland, Kansas. Two Curtiss 7hp V-twin aeronautical engines were purchased from the Curtiss Company for the helicopter midyear and demonstrated to the public in November. Purvis and Wilson would receive a patent for the helicopter later in 1912, making it the first patented rotary-winged aircraft in the United States.

A major change took place in the finish of the motorcycles this year. The only color now offered was "Curtiss Silver Grey," with black as an unadvertised option.[4] This would remain the standard through 1912, when motorcycles were discontinued.[5]

While in Bedeck, Nova Scotia, working with Bell in February, Glenn helped construct an iceboat powered by a Curtiss 15hp, four-cylinder, in-line engine. During a test run on the frozen lake there, Glenn, while going an estimated 60 mph, was lax in slowing down the sled and as a result ran into a shed on the shoreline.[6] While he wasn't seriously injured, it was here that he received his trademark scar below his lower lip. As a result, Glenn would very rarely smile for photographs, since the scar made it painful.

Following Glenn's success with Bell's Aerial Experiment Association, the G. H. Curtiss Manufacturing Company was reorganized in March under the new name Herring-Curtiss Company. A pioneer aeronaut, Augustus Herring joined Curtiss as a business partner with an emphasis on "development and manufacture of aeroplanes." The new firm promised the public that in no way would it affect the Curtiss-brand motorcycle manufacture.[7] To back this up, the company would build two new two-story buildings "exclusively for

One of two Curtiss V-Twin engines used in the 1909 Purvis-Wilson helicopter experiments. *Photo: Richard Leisenring Jr.*

motorcycle manufacture."[8] Advertising the Curtiss motorcycle under the new company name would immediately start in April with the issuing of a Herring-Curtiss motorcycle catalog for the following year.

At this time, the Herring-Curtiss Company built and sold an airplane named the "Golden Flier" to the newly formed Aeronautic Society of New York. This would be the first commercial sale of an airplane in the United States.

To bolster the stability of the motorcycle end of the business, advertisements for the Curtiss motorcycles would point out the laurels won by independent racers on the machines, such as a perfect score by the only Curtiss cycle in the 1909 New York Endurance Run and the three perfect scores set in the California Endurance Run, two on twin-cylinder cycles—the only two-cylinder machines to finish the event.[9]

Glenn left in August for Rheims, France, to participate in the First International Aviation Meet, representing the Aero Club of America. Using a quickly built aircraft named "the Rheims Racer," he successfully won the coveted Gordon Bennett Silver Cup (valued at $2500) and a total of $7,600 in prizes. During the competition, Glenn sprained an ankle, requiring the use of a cane for a short time to get around.

Proceeding to Brescia, Italy, he won another $7,000 in prizes participating in their air meet. The air meets offered not only prestige and magnificent trophies but also enormous cash prizes unheard of in motorcycle competitions. It was at this time that the Wright brothers began patent infringement lawsuits against anyone building and selling aircraft—specifically, the Herring-Curtiss Company.

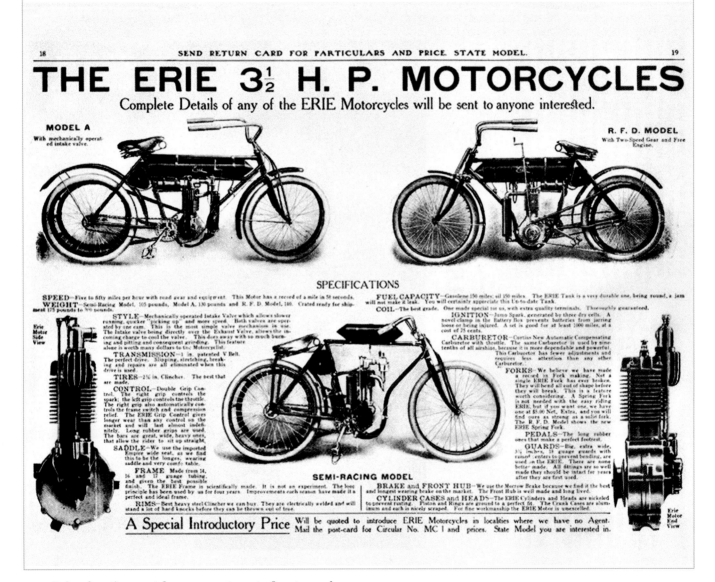

1909 Erie advertisement for an assortment of motorcycles

This would be the last year the Erie would be produced as a complete motorcycle, thereafter becoming a kit instead, with assembled versions being phased out in late 1910. Tank was now heading into new ventures as well as arranging for the construction of new buildings for the Motorcycle Equipment and Supply Company.[10]

Tank would also operate for a brief time—in conjunction with the Motorcycle and Equipment Supply Company (MESCO)—a business called the Bicycle Motor Company (BMC) to sell engine castings around 1909–10. However, very little is known of BMC other than a few small advertisements. Tank reorganized

1905 Erie motorcycle. *Courtesy of the Barber Motorsports Museum, photo by Derek Snyder*

MESCO under a new name, the Motorcycle Equipment Company (MECO), with Glenn Curtiss as an investor and incorporator.[11]

A new and unusual motorcycle was offered this year under the Erie brand name—the 1909 Erie Railroad Motorcycle. This unique machine utilized four altered tire rims to fit railroad rails and was designed to run on standard-size railroad tracks while powered with a choice of either a 3.5hp or 7hp Curtiss single-cylinder engine.[12] Originally designed for railroad workers checking lines, it was also advertised as a great individual transport to hop from one town to another. This was actually another version of a tricycle railroad motorcycle built by the Curtiss Company in 1907 but never put into production.[13]

Motorcycle Equipment Company postcard advertisement

Erie Railroad Cycle

The Curtiss factory, taken from the hill behind the facility, with the village of Hammondsport in the background, 1909

CHAPTER 14

THE MARVEL

Marvel logo. *Illustration: Joshua Leisenring*

In August 1909, Glenn formed yet another enterprise, the Marvel Motorcycle Company, in partnership with and under the leadership of Tank Waters. Glenn would hold 49 percent of the stock, while Tank held 50 percent and Elizabeth Waters one share. A new building to house Marvel was erected on property adjacent to the Curtiss plant, with the intention of employing a large workforce. Lena Curtiss would later become the property deed holder.[1] This was done to protect the interests from possibly being used in the Wright aviation patent lawsuit. While Marvel would produce only motorcycles, it would contract with the Herring-Curtiss Company to produce five hundred engines at $35 apiece and advertise them as "Curtiss" in order to keep the two companies separate, yet to capitalize on the Curtiss name.[2]

In conjunction with the new brand of motorcycle, an innovative new engine was offered this year. A 5hp, one-cylinder, overhead-valve, air-cooled engine was designed with the combined talents of Curtiss employees Alexander Pfitzner and Henry Kleckler. Having developed a prototype during the winter of 1908–09, Henry installed it on his personal motorcycle and road-tested it to prove its durability during the summer of 1909.[3] As soon as Kleckler felt the engine had sufficient testing, the job of producing working blueprints and drawings of all the necessary parts was turned over to Curtiss employee Robert Patterson in the fall.[4] Known as the Model G and used both on the new Marvel and Curtiss machines, the one-cylinder engine proved to be as powerful as the older-style V-Twin, rating in at 5 hp, essentially making the V-Twin obsolete. Also briefly offered in 1910 was an additional 9hp, double-cylinder, overhead-valve version (no model number) strictly for the Curtiss brand.[5] Whether any were actually produced or sold commercially is unknown at this time.

Curtiss Model G engine

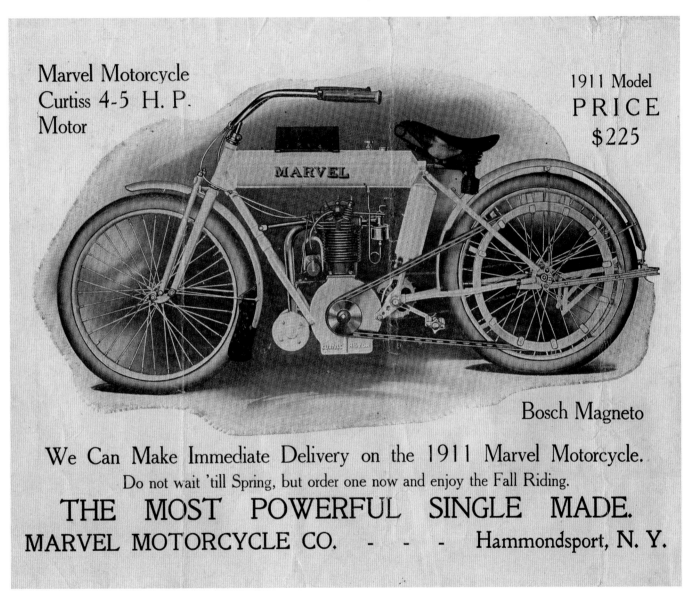

Advance advertising sheet for the Marvel, 1910

The Marvel itself was unique in that its oil and gas tanks were an integral part of the frame. Advertised as "Tanks-in-the-Tube," it was designed by former MESCO and now Marvel employee Clarence P. Rudd, with input from Curtiss employee William T. Thomas. This concept was a radical departure from the various bicycle-style frames offered by Curtiss and other companies over the years. Unlike the Curtiss cycles, which were offered in "Curtiss Silver Grey," the Marvel finish was a much-lighter "Marvel Grey."[6]

Unfortunately, the Marvel Company got off to a bad start. Production of the cycles was behind schedule due to Herring-Curtiss's inability to supply the contracted engines on time. This was a result of a parts shortage and various other delays caused by the Herring-Curtiss Company (delayed submission of parts drawings to contractors) and outside suppliers.[7] The reason behind many of the Model G parts being outsourced instead of being made in-house as traditionally done was due to the increase in the manufacture of

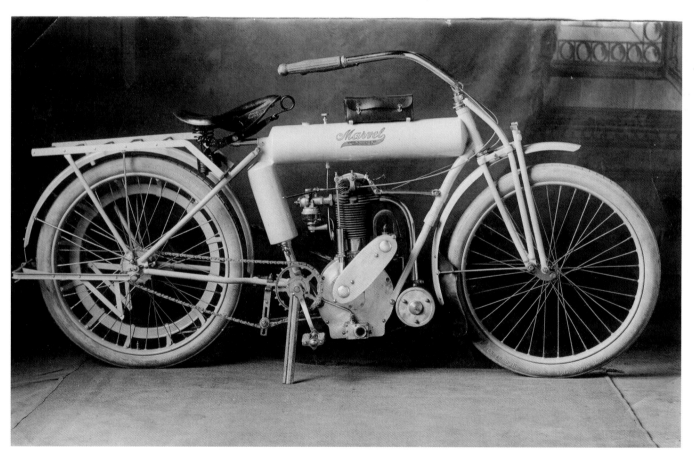

Advertising photo of the 1911 Marvel

aeronautical engines. The projected number of 1,500 motorcycle engines (1,000 for Curtiss) would have left the company unable to meet the aeronautic-engine demands.[8]

The 1910 Marvel motorcycle would make its public debut at the New York Auto Show in early January.[9]

However, because of the parts shortage, the model displayed at the show was equipped with a dummy engine. Enough parts to assemble a few engines would finally arrive in late May to early June, to complete a very small number of 1910 model cycles[10]—much too late for the traditional sales season.

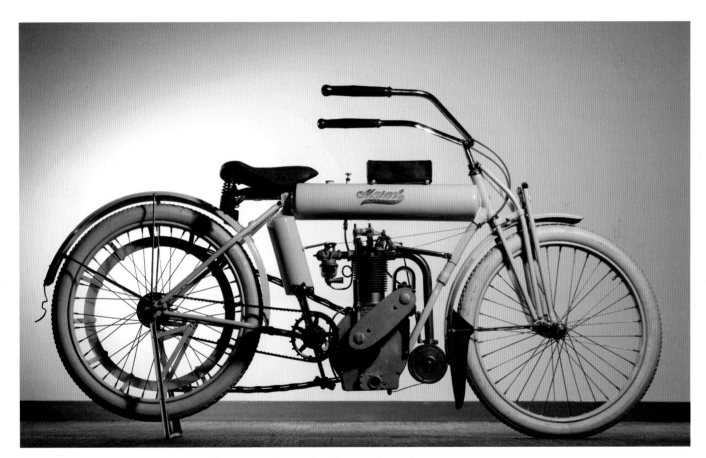

1910 Marvel motorcycle. *Courtesy of Bruce Lindsay; photo by Derek Snyder*

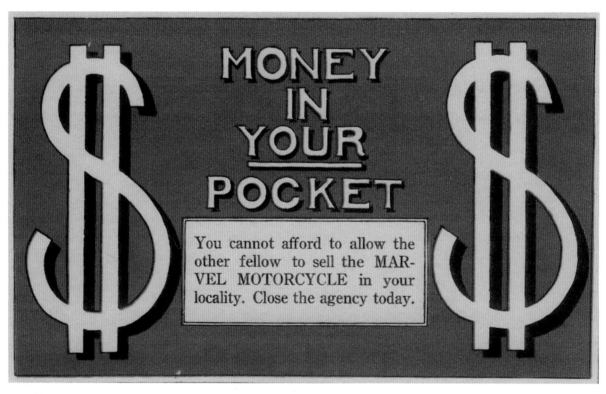

Advertising postcard soliciting agents for the Marvel motorcycle

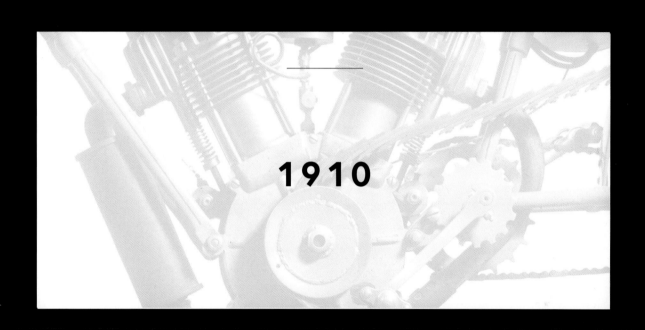

CHAPTER 15

1910

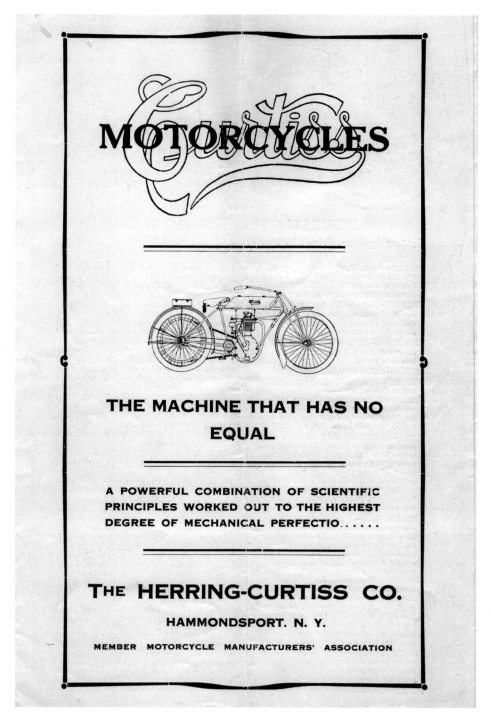

Cover of 1910 Herring-Curtiss advance sheet

Sadly, 1910 heralded the demise of the Curtiss motor-cycle and the industry as a whole in Hammondsport. The Curtiss-Herring partnership quickly soured and soon dissolved, with "little chance for the Curtiss plant being opened as a motorcycle factory," according to one news report.

When Glenn went into the partnership with Augustus Herring, it was with the agreement that Herring would put up patents he claimed to have that would negate the Wright brothers' lawsuit and would allow the Herring-Curtiss Company to continue to produce aircraft. Unfortunately, this was not the case, since Herring did not possess patents of that nature to hold up his end of the contract. In order to dissolve the partnership, Glenn allowed the company to go into bankruptcy. Herring would launch a lawsuit to gain control of the company. In the meantime, Glenn competed in May for the $10,000 prize offered by *New York World* for the first person to fly the length of the Hudson River from Albany to New York City, easily taking the prize.

At this point, it was estimated that the company produced three hundred motorcycles.[1] The company was now controlled by George H. Parkhurst as the court-appointed receiver. With the plant idle due to impending bankruptcy and lawsuits because of the split, Glenn managed to rent a portion of the facility back from the receivership. This kept a small staff employed between the months of April and December.[2] The goal was to fill existing orders for Curtiss motor-cycles, using parts still on hand at the factory, which Glenn also purchased from the receivership. The Model G motorcycle engine, while a few may have been pro-duced at the Herring-Curtiss plant, was continued at the Marvel plant for both brands. Parts suppliers began forwarding parts for engine production directly to Marvel instead of Herring-Curtiss, in an attempt to keep the Curtiss motorcycle brand name alive and keep the Marvel Company from folding.[3] Glenn would also turn over his share of Marvel stock to Waters to protect the company should the Herring lawsuit spill over on it—which it did.[4] A dispute would arise as to settling of accounts during the subsequent bankruptcy proceedings of the Herring-Curtiss Company.

An interesting sidenote regarding the 1910 Curtiss-model motorcycles is how they were serial numbered. According to Harry Genung during his testimony at the 1922 Herring-Curtiss lawsuit, serial numbers for this year started at 6000. An example of this was a receipt presented as evidence that read "F. Ellis, Vestal, New York, one second hand 1910 single cylinder machine, no. 6002, $120." Hence, 6002 was the third cycle made for that model year. He also stated that numbering was changed each year so that there would not be any confusion in the orders received.[5] Many of the Model G engines would be found with a variety of markings, causing confusion as to when or where they were made. It is also noted that on many of the examined casings, there can be found "G10." For many years, it was thought to stand for the model and year, when in fact it is the part number, which, in this instance, number 10 is the casing.[6]

Over the course of the years, Curtiss would offer the public a wide variety of machines both stock and custom built, from touring bikes to racers, as well as frame styles and accessories. Curtiss engines could be had in different cylinder configurations, strokes, and cycles, as well as various horsepower, depending on their intended use. Whatever the public wanted, the Curtiss Company, regardless of the name it oper-ated under, strove to provide, and for many years, Curtiss motorcycles were considered "the greatest roadsters on earth."

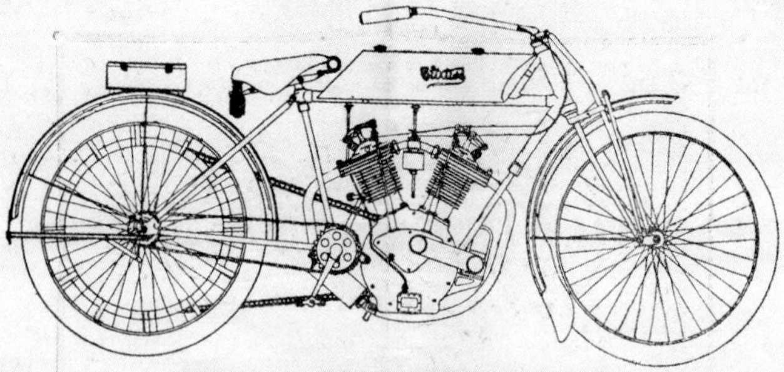

DOUBLE CYLINDER, 8-9 H. P.

Battery Ignition, Price $275 **Magneto Ignition, Price $300**

SPECIFICATIONS

MOTOR—8-9 H. P., roller bearing, 3¼ in. bore, 3⅝ in stroke, valves in the head, the latest and best type of construction, positively vibrationless. The smoothest running motor in the world.

VALVES—Exceptionally large, 1 11-16 in. diameter, interchangeable, made from imported steel, specially treated, mechanically operated by one push rod from one cam.

BEARINGS—Roller on main shaft and lower end of connecting rods, hollow steel piston pins with bearing in piston.

SPEED—4 to 65 miles per hour. (Guaranteed)

WEIGHT—Exceptionally light, 150 pounds with full road equipment.

FRAME—17 inches high. (Lowest saddle position in the world.) 1¼ in. Shelby seamless tubing, heavy ¾ in. forks and stays, all parts reinforced. Drop forged rear fork tip, etc.

TANK—Special heavy block tin with reinforcement to insure quietness. The gasoline outlet is fitted with special strainer readily removable for cleaning purposes. Capacity 2 gallons gasoline, 2 quarts oil.

TRANSMISSION—Curtiss Perfected 28 degree 1⅛ inch V Belt made from specially prepared leather. Will not slip, stretch or break. The best transmission in the world for motorcycles.

LUBRICATION—The simplest and best system ever incorporated in a motorcycle. Oil well in the crank case provided with a float which automatically shuts off the supply when it reaches the proper level. The correct level is set at the factory and requires no further attention other than to replenish the supply.

IGNITION—2 standard dry cells and special double coil in strong steel case. This case is very neat and compact and is positively rain and dust proof. Short circuits are practically impossible. The switch is also built inside this case. **Option:** Curtiss high tension magneto driven by ¼ inch roller chain and sprockets enclosed in oil tight case, $25.00 extra.

CONTROL—Double grip. Right grip, circuit spark advance and compression. Left grip throttle. Our system of control is very simple and positive. The grips stay in any desired position.

FORKS—Curtiss improved spring fork. Very liberal bearings. A fork that overcomes all road irregularities without jolt or jar. Can be changed to rigid fork in 5 minutes.

CARBURETOR—Is of the most approved central draft type. Adjustments on both needle and auxillary air valves. Very light, neat and compact. Throttles perfectly at all speeds. Our special gasoline shut-off prevents flooding and shaking up. The fastest and most powerful carburetor in the world.

HANDLEBARS—V shaped. Very long and comfortable. Strongly reinforced inside and out. Securely fastened by both compression clamp and expander. Rough rider grips.

FINISH—Curtiss "Silver Grey." All bright parts nickled on copper. Particular attention is given to the finishing of our machines, each coat of enamel being baked and rubbed separately.

WHEELS—28 inch. 36 10 gauge spokes, front and rear. Knockout front hub. Eclipse Coaster Brake especially designed for Curtiss Motorcycles.

RIMS—Best grade of steel. Heavily coppered to prevent rust.

TIRES—2 1-2 inch Kokomo Gridiron tread. Bailey Treads extra $1.50 per pair.

MUFFLER—The most effective sound silencer in the world. No back pressure. Large expansion chamber. The quietness of our machines is one of their chief charcteristics. Cut-out operated by foot.

GEAR—3 1-2 to 1. Optional 3 or 2 1-2 to 1.

MUD GUARDS—18 gauge. Extra wide. Double braces in rear. Front guard fitted with large mud splasher.

SADDLE—Persons Champion Motor Seat.

FRONT PULLEY—Hardened steel. Deeply grooved to insure long life of belt.

REAR PULLEY—Pressed steel. Attached to the rim by 12 strong braces.

STAND—Curtiss "Special." Operated independent of axle. Swings up and attaches to mud guard when not in use.

SPARK PLUG—Curtiss "Special." Guaranteed against porcelain breakage.

TOOL BAG—Persons (tourist type.) Fastens to top of rear mud guard. Contains full kit of tools for engine and machine.

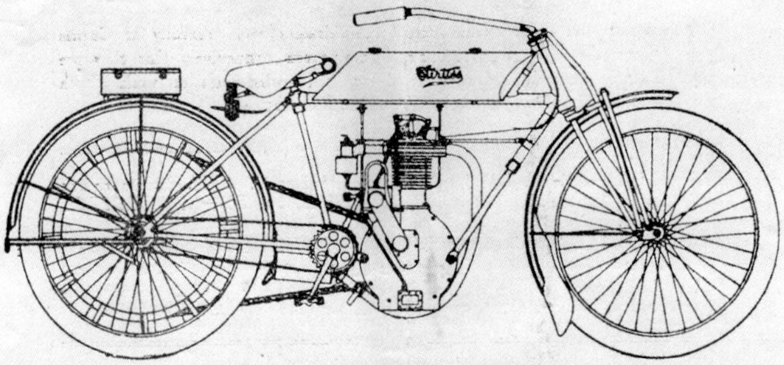

SINGLE CYLINDER, 4-5 H. P.

Battery Ignition, Price $200 **Magneto Ignition, Price $225**

SPECIFICATIONS

MOTOR—4-5 H. P., roller bearing, 3¼ in. bore, 3⅝ in stroke, valves in head, most advanced type of construction, off-set cylinder, silent and absolutely vibrationless.

VALVES—1 11-16 in. diameter, interchangeable, mechanically operated by one push rod from one cam. Made from imported steel, specially treated.

BEARINGS—Roller on main shaft and lower end connecting rod. Piston pin held in rod and bears in bronze in piston.

SPEED—4 to 55 miles an hour. (Guaranteed.)

WEIGHT—125 pounds with full road equipment.

FRAME—17 inches high (lowest saddle position in the world.) Shelby seamless tubing, 1¼ in. 14 gauge, strongly reinforced. Rear fork tip, etc. drop forgings.

TANK—Extra heavy block tin with zig-zag reinforcement to insure quietness. Large gasoline strainer in tank. Capacity 2 gallons gasoline, 2 quarts oil.

TRANSMISSION—Curtiss perfected 1⅛ in., 28 degree V belt, made from specially prepared leather, guaranteed not to slip, stretch or break. Generally acknowledged to be the best transmission in the world for motorcycles.

LUBRICATION—The simplest and best system ever incorporated in a motorcycle. Oil well in crank case provided with a float which automatically shuts off the supply when it reaches the proper level. The correct level is set at the factory and requires no further attention other than to replenish the supply.

IGNITION—2 standard dry cells and special coil in strong steel case. Very neat and compact. Absolutely water proof. The switch is also built inside the case and is thus protected from the dirt and dust. **Option**—Curtiss magneto, driven by ¼ in. roller chain and sprockets enclosed in oil tight case. $25 extra.

CONTROL—Right grip controls circuit, spark advance and compression relief. Left grip, throttle. Our system of control is very simple and positive. The grips always stay where you put them.

FORKS—Latest improved spring fork, strongly made, very liberal bearings, can be changed to rigid type in 5 minutes. Absorbs all road shocks and jars.

CARBURETOR—Is of the most approved central draft type. Adjustments on both needle and auxillary air valves. Very light, neat and compact. Throttles perfectly at all speeds. Special gasoline shut-off prevents flooding and shaking up on rough roads. We have demonstrated that we can get more power and speed from this carburetor than any other in the world.

HANDLEBARS—V shaped, very long and comfortable, strongly reinforced inside and out, fastened by both compression clamp and expander. Rough Rider Grips.

FINISH—Curtiss " Silver Grey." Great care is taken in the finishing of our machines, each coat of enamel being baked and rubbed separately.

WHEELS—28 in., 36 spokes front and rear. Knock-out front hub. Special heavy coaster brake.

RIMS—Best grade of steel, heavily coppered to prevent rust.

TIRES—2½ in. Kokomo detachable. Bailey treads extra, $1.50 per pair.

MUFFLER—Most effective in the world. Large expansion chamber, absolutely quiet and no back pressure. The quietness of our machines is one of their distinctive features. Efficient cut-out operated by foot.

GEAR—4 to 1. Optional 4½ or 3½ to 1.

MUD GUARDS—18 gauge, very wide, double braces in rear. Front guard fitted with large splasher.

SADDLE—Persons Champion Motor Seat.

FRONT PULLEY—Hardened steel. Very deep groove to insure long life of belt.

REAR PULLEY—Pressed steel attached to rim by 12 strong braces.

STAND—Special. Operated independent of axle. Swings up and fastens to mud guard when not in use.

SPARK PLUG—Curtiss special. Guaranteed against porcelain breakage.

TOOL BAG—Persons. (Tourist type) Attached to top of rear mud guard. Contains full kit of tools for engine and machine.

1910 Herring-Curtiss advance sheet advertising the V-Twin and the Model G on Wehman frames

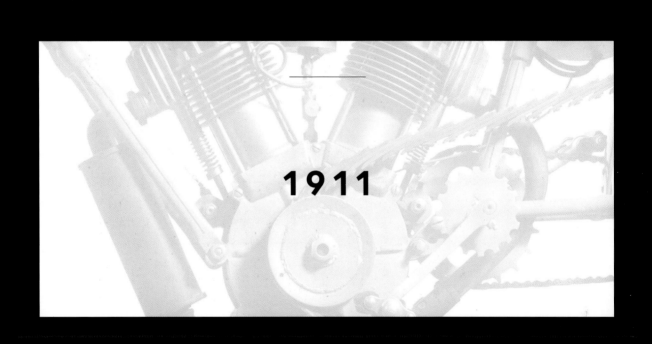

CHAPTER 16

1911

The Curtiss
factory
buildings, ca.
1911

Another view
of the Curtiss
factory
buildings, ca.
1911

1911 Curtiss Cook motorcycle with Model G engine. *Photo: Derek Snyder*

1911 Curtiss Motorcycle Company catalog cover

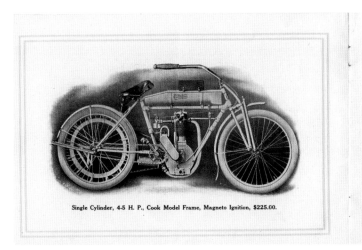

Single Cylinder, 4-5 H. P., Cook Model Frame, Magneto Ignition, $225.00.

1911 catalog featuring the Curtiss Cook motorcycle

The year 1911 saw many changes at the manufacturing facilities in Hammondsport. After the bankruptcy of the Herring-Curtiss Company, Curtiss purchased back the assets of the company. The Curtiss Aeroplane Company was formed in April, and the Curtiss Motor Company a few months later to take charge of the existing plant (the two would merge in 1914 to form the Curtiss Aeroplane & Motor Company). The new unincorporated Curtiss Motorcycle Company that appeared in May was merged with the Marvel Company by December.

In the meantime, the 1911 model of Marvel and Curtiss motorcycles again appeared on the market as one-cylinder, overhead-valve machines, with the Marvel Company advertising that the Model G was being made in their facility.[1] The Curtiss 1911 catalog offered both Wehman and standard-model frames, the V-Twin being completely discontinued. An "O"-model Curtiss cycle with the older-style 3.5hp, one-cylinder engine on a standard frame was also offered to use up existing surplus stock. The Marvel would see two slight changes in that it would use a Heitger glass bowl carburetor instead of a Curtiss carburetor, and a Curtiss front fork instead of the simpler Sager-style fork designed by J. Harry Sager of Rochester, New York. The Marvel logo would also go from block letters to a script, similar to the Curtiss.[2] The Marvel would use a small decal on the front fork instead of a head badge.

One addition to the 1911 catalog was the introduction of the Curtiss Cook motorcycle. Weighing in at a hefty 175 pounds and utilizing the new Curtiss Model G, single-cylinder, overhead-valve engine, the cycle was designed by Albert Cook for endurance races that he himself excelled in as a representative for the Curtiss Company. This would be just the second time that the company would offer a stock racing cycle in a catalog.

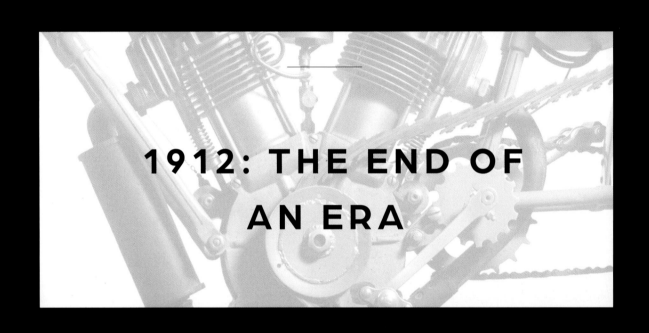

CHAPTER 17

1912: THE END OF AN ERA

1912 Curtiss motorcycle. Wehman frame with the Model G engine. *Photo: Derek Snyder*

Sales quickly diminished (due in part to the lack of variety, the rumored instability of the company, and stiff competition in the growing industry), and 1912 would be the last year that Curtiss and Marvel motorcycle and motorcycle engine production of any kind would take place in Hammondsport. It was reported that all the major contributors of the Marvel and Curtiss cycle interests had moved on to other ventures by November. The major emphasis was now fully on aviation in the village.

Hammondsport-built motorcycles were, as of 1912, a thing of the past, with no sign of them ever being resurrected. As a final blow, the "Curtiss Motorcycle Company" announced in *Motorcycling Magazine* the following year, on November 24, 1913—as well as in several other publications—that "owing to the rapid increase of our aeronautical business . . . we offer for immediate sale our complete stock of motorcycle parts, including designs, jigs, tools, goodwill, etc." Tank Waters would continue operation of the Motorcycle Equipment Company (MECO) in Hammondsport through 1929 strictly as an accessory supply company, with a West Coast branch in Los Angeles, California, from 1914 to 1920. And while there were Harley-Davidson and Excelsior dealerships in the village for a short time in the 1920s and '30s, any connection the village had to the motorcycle industry ceased to exist in its entirety.

As for dirigibles, the Curtiss Aeroplane & Motor Company would produce nineteen semirigid dirigibles during the early part of World War I. One of those set a nonstop endurance record of thirty-two hours on November 23–25, 1918—a great advancement from the four-hour record that Glenn had set exactly eleven years earlier, to the day, in 1907.[1] Three companies would operate out of Hammondsport for a short period after the war. These were the Meadowcroft Balloon Works; Airships, Inc.; and Air Cruisers, Inc., all occupying the former Curtiss factory buildings.

The year 1912 would also have a bright side in Hammondsport, with Glenn and Lena welcoming into the world their second son, Glenn Jr., on June 16, 1912, ten years and four months after the loss of their first son, Carleton. With a new family and new industry, Glenn would find himself in a completely new phase of his life.

While Glenn H. Curtiss and Hammondsport are famous for their major influence in aviation history, it should be noted that when it comes to the motorcycle industry, engine development, and the early dirigible era, Glenn's impact is just as great. For the ten short years from 1902 to 1912, Glenn and his representatives set world speed records, won countless races, and inspired other companies to move on to bigger and better accomplishments. Five brands of motorcycles were produced in Hammondsport—the Hercules, the Curtiss, the Erie, the MECO, and the Marvel. The Curtiss motorcycle would be produced under the G. H. Curtiss Manufacturing Company, the Herring-Curtiss Company, and finally the Curtiss Motorcycle Company.[2]

No doubt, had Glenn Hammond Curtiss kept his interest in motorcycles on an equal par with his aircraft, several famous brands might not be where they are today. And as it once was noted, "If it wasn't for his motorcycles, the airplanes would have never gotten off the ground."[3]

The Motorcycle Equipment Company building on Lake Street, Hammondsport, New York, ca. 1912

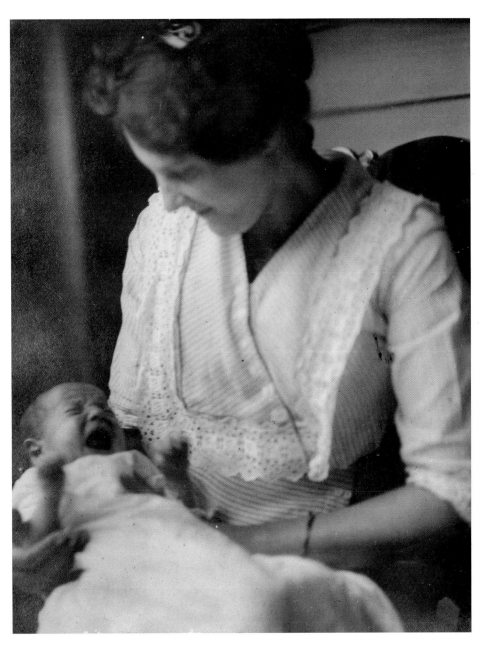

Lena with Glenn Jr.

List of Known Motorcycle Events That Glenn H. Curtiss Personally Attended or Participated In

1902

September 1: Brooklyn, New York, New York Motorcycle Club
10-Mile Handicap, 3rd-Place Medal
Overall Time, 2nd-Place Trophy

1903

May 30: Brooklyn, New York, New York Motorcycle Club
Riverdale Hill Climb, 1st-Place Medal
May 30: Yonkers, New York, National Cycle Association
1-Mile Race, medal & world speed record
5-Mile Race, 1st-Place Medal
September 5–7: Brooklyn, New York, Federation of American Motorcyclists Meet
Registered but no evidence of attending
September 12: Syracuse, New York, New York State Fair
No wins or places

1904

January 29–30: Ormond Beach, Florida, speed trials
1-Mile Race, 1st-Place Trophy & American speed record
10-Mile Race, 1st-Place Trophy & world speed record
May 30: Brooklyn, New York, New York Motorcycle Club

Riverdale Hill Climb, 2nd Place
July 4: New York to Cambridge, Maryland, Federation of American Motorcyclists endurance races
No wins or places
September 5: Tonawanda, New York
5-Mile Open, 1st Place
5-Mile Exhibition, 1st Place
15-Mile Open, 2nd Place
November 8: Hackensack, New Jersey, North Jersey Motorcycle Club, Election Day Races
Attended as a spectator

1905

July 4: Chicago, Illinois, Chicago Motorcycle Club
Exhibition Run attended to demonstrate V-Twin
August 8: New York to Waltham, Massachusetts, Federation of American Motorcyclists Endurance Races
25-Mile Race, 1st-Place Medal
25-Mile Race, Time Prize Trophy
September 18: Syracuse, New York, New York State Fair
5-Mile, One-Cylinder Race, 1st-Place Trophy & world speed record
5-Mile, Two-Cylinder Race, 1st-Place Trophy & world speed record
3-Mile, One-Cylinder Race, 1st-Place Trophy

1906
July 4: Rochester, New York, Motorcycle Races
3-Mile Race, 1st Place but later disqualified for weight limit
5-Mile Race, 1st Place but later disqualified for weight limit

1907
January 22–25: Ormond Beach, Florida, speed trials
1-Mile, One-Cylinder Race, 1st-Place Medal & world speed record
1-Mile, Two-Cylinder Race, 1st-Place Medal & world speed record
Time Trial, V-8 motorcycle, world land speed record
May 30: Manhasset Hills, Long Island, New York, New York & Brooklyn Motorcycle Club
Class III Event, 1st Place
Free-for-All Event, 1st Place
June 27: Hammondsport, New York, Gala Day Race
Sponsored by Glenn H. Curtiss

July 31–August 3: Providence, Rhode Island, Federation of American Motorcyclists endurance races
Free-for-All Hill Climb, One-Cylinder, 1st-Place Trophy
Free-for-All Hill Climb, Two-Cylinder, 2nd-Place Trophy
1-Mile Time, One-Cylinder Race, 1st-Place Trophy & world record
1-Mile Time, Two-Cylinder Race, 1st-Place Trophy

1908
June 6: New York City, New York, New York Motorcycle Club, Fort George Hill Climb
Free-for-All, 2nd Place
Class B, One-Cylinder, 3rd Place
Class D, Two-Cylinder, 4th Place
October 24: Long Island Vanderbilt Cup Race
Participated as a volunteer motorcycle courier

APPENDIX II

List of Known Early Curtiss Employees from 1900 to 1910

Bailey, J.
Bauter, E. P.
Bauter, Lynn
Blevin, Floyd
Brandow, David
Brown, Byron
Carroll, Sylvester
Clark, Eddie
Clark, William
Cook, Albert D.
Curtiss, Lena
Damoth, William "Will"
Davis, Dee James
Dillenbeck, Walter
Dorsey, S.
Douglas, Jay
Feagle, Flint Barney
Genung, Harry
Hollenbeck, H.
Horton, H.
Hunt, Thomas
Hunt, Wheeler
Kleckler, Henry

Longwell, Harry
Losey, L.
Losey, Tim
Lovell, F.
Lovell, W.
Lowry, Lucian
Mackey, G.
Mahan, Frank
Miller, Claude
Morrow, J. E.
Osborne, George
Osborne, John "Jack"
Patterson, Robert
Pfitzner, Alexander
Safford, Jay
Smith, F. A.
Sutterly, L.
Swarthout, Anna
Taylor, Frank
Thomas, William T.
Wakeman, Charles
Wehman, Henry J. "Hank"
White, Clarence
Willard, C.
Wixom, Charlie

NOTES

Chapter 1: The Early Days

1. *Rochester (NY) Democrat & Chronicle*, June 7, 1892.
2. "Curtiss, Rutha," Rutha Curtiss Memories (unpublished document), Curtiss, Rutha, file #574.0, G. H. Curtiss Museum, Hammondsport, NY.
3. Ibid.
4. Ibid.
5. Jas. Smellie letter to E. C. Stearns & Co., May 31, 1897, Curtiss Bicycles file #609.0, G. H. Curtiss Museum, Hammondsport, NY.
6. *Hammondsport (NY) Herald*, May 30, 1900.
7. "Curtiss, Rutha," Rutha Curtiss Memories (unpublished document), Curtiss, Rutha, file #574.0, G. H. Curtiss Museum, Hammondsport, NY
8. Ibid.
9. *Hammondsport (NY) Herald*, March 29, 1899.
10. *Hammondsport (NY) Herald*, March 31, 1900.

Chapter 2: 1900

1. *Rochester (NY) Democrat & Chronicle*, October 19, 1900, 4.
2. Ibid.
3. E. R. Thomas Motor Co.'s Products, in *Horseless Age*, October 10, 1900, 17; and Castings of Thomas Motors, in *Horseless Age*, November 7, 1900, 82.
4. *Hammondsport (NY) Herald*, November 14, 1900.

Chapter 3: 1901

1. *Hammondsport (NY) Herald*, January 22, 1902.
2. "Curtiss, Rutha," Rutha Curtiss Memories (unpublished document), Curtiss, Rutha file #574.0, G. H. Curtiss Museum, Hammondsport, NY
3. Ibid.
4. *Hammondsport (NY) Herald*, May 4, 1904.

Chapter 4: 1902

1. *Hammondsport (NY) Herald*, June 18, 1902.
2. *Hammondsport (NY) Herald*, March 12, 1902.
3. *Hammondsport (NY) Herald*, August 22, 1902; and Letter of recommendation, June 4, 1906, Feagles, Flint Barney file #182.0, G. H. Curtiss Museum, Hammondsport, NY.
4. *Hammondsport (NY) Herald*, October 15, 1902.
5. *1903 Hercules Motor Cycles & Motors* (catalog) (Hammondsport, NY: G. H. Curtiss Mfg., 1903).
6. *Hammondsport (NY) Herald*, September 3, 1902; and *Hammondsport (NY) Herald*, September 10, 1902.
7. *Hammondsport (NY) Herald*, December 3, 1902.

Chapter 5: 1903

1. *Hammondsport (NY) Herald*, April 1, 1903.
2. *Hammondsport (NY) Herald*, April 26, 1905.
3. *Hammondsport (NY) Herald*, April 22, 1903.
4. *Hammondsport (NY) Herald*, October 15, 1902.
5. Geoffrey N. Stein, *The Motorcycle Industry in New York State* (Albany: University of the State of New York, 2001), 12.
6. *Hammondsport (NY) Herald*, January 14, 1903.
7. *Hammondsport (NY) Herald*, April 15, 1903.
8. *Hammondsport (NY) Herald*, January 4, 1905.
9. "Broke the World's Record: This Is the Man and the Machine Did It," unknown publication, 1903, Clara Studer Collection, R-1-51, National Air Museum, Smithsonian Institution, Washington, DC.
10. *1903 Hercules Motor Cycles & Motors* (catalog) (Hammondsport, NY: G. H. Curtiss Mfg., 1903).
11. *Hammondsport (NY) Herald*, March 18, 1903.
12. *Hammondsport (NY) Herald*, July 8, 1903.
13. *Hammondsport (NY) Herald*, July 15, 1903.
14. *1904 Motorcycles and Motors* (catalog) (Hammondsport, NY: G. H. Curtiss Mfg., 1904).
15. *5th Annual Catalogue Curtiss Motorcycles and Motors* (Hammondsport, NY: G. H. Curtiss Mfg., 1906).
16. *Hammondsport (NY) Herald*, May 6, 1903.
17. *Hammondsport (NY) Herald*, April 22, 1903.
18. *Horseless Age*, June 3, 1903; *Star-Gazette* (Elmira, NY), June 1, 1903, 6; and *Hammondsport (NY) Herald*, June 3, 1903.
19. "Motor Cycle Contest," *Hartford (CT) Courant*, July 4, 1903, 13.
20. *Hammondsport (NY) Herald*, September 16, 1903.
21. 1904 catalog (St. Louis, MO: Harry R. Geer, 1904).
22. *Hammondsport (NY) Herald*, October 21, 1903.
23. *Hammondsport (NY) Herald*, August 3, 1904; and "Propelled by a Curtiss Motor, Only Successful Airship at St. Louis," *Hammondsport (NY) Herald*, November 30, 1904.

Chapter 6: 1904

1. "Hammondsport Man Made Records," *Hammondsport (NY) Herald*, February 6, 1904.
2. *Hammondsport (NY) Herald,* May 11, 1904.
3. *Bicycling World & Motorcycle Review*, June 1904, 322.
4. *Hammondsport (NY) Herald*, July 18, 1906.
5. *Hammondsport (NY) Herald*, July 14, 1904.
6. *Hammondsport (NY) Herald*, July 27, 1904.
7. *Hammondsport (NY) Herald*, July 20, 1904.
8. *Hammondsport (NY) Herald*, November 9, 1904.
9. "MOTOR CYCLISTS," *The Record* (Hackensack, NJ), October 3, 1904, 1.
10. *Hammondsport (NY) Herald*, May 18, 1904; and *Hammondsport (NY) Herald*, October 4, 1905.
11. *Hammondsport (NY) Herald*, May 25, 1904.
12. *Hammondsport (NY) Herald*, September 7, 1904.
13. *Hammondsport (NY) Herald*, May 25, 1904.
14. *Hammondsport (NY) Herald*, June 8, 1904.
15. "Hercules Gives Way To Curtiss," *Bicycling World & Motorcycle Review*, February 15, 1905, 1.
16. *1904 Motorcycles and Motors* (catalog) (Hammondsport, NY: G. H. Curtiss Mfg., 1904).
17. Ibid.; and "Two New Motor Cycles," *Scientific American Magazine*, February 20, 1904, 159.
18. "Will Champion Curtiss Fly in Chicago?," *Chicago Tribune*, October 10, 1909, 66.
19. "Propelled by a Curtiss Motor; Only Successful Airship at St. Louis," *Hammondsport (NY) Herald*, November 30, 1904.
20. Stein, *The Motorcycle Industry in New York State*, 23.

Chapter 7: 1905

1. *Hammondsport (NY) Herald*, January 25, 1905.
2. "Curtiss, Rutha," Rutha Curtiss Memories (unpublished document), Curtiss, Rutha file #574.0, G. H. Curtiss Museum, Hammondsport, NY.

3. *Hammondsport (NY) Herald*, May 3, 1905.
4. *Chicago Tribune*, July 2, 1905.
5. *Hammondsport (NY) Herald*, November 22, 1905.
6. *Cycle and Automobile Trade Journal*, 1905.
7. *Hammondsport (NY) Herald*, October 4, 1905; *Annual Catalogue 1905 Curtiss Motorcycles and Motors* (Hammondsport, NY: G. H. Curtiss Mfg., 1905); and G. H. Curtiss Mfg. Co. Incorporation, October 17, 1905, Curtiss Collection, box 10, folder 1, National Air & Space Museum. Washington, DC.
8. *Waterbury (CT) Democrat*, August 10, 1905.
9. *Burlington (VT) Free Press*, September 19, 1905.
10. *Hammondsport (NY) Herald*, November 8, 1905.

Chapter 8: 1906

1. *Hammondsport (NY) Herald*, January 10, 1906.
2. *Hammondsport (NY) Herald*, March 7, 1906.
3. MESCO Order Card, 1906, copy, Waters, C. Leonard file #531.0, G. H. Curtiss Museum, Hammondsport, NY.
4. *5th Annual Catalogue Curtiss Motorcycles and Motors* (Hammondsport, NY: G. H. Curtiss Mfg., 1906).
5. *Hammondsport (NY) Herald*, June 27, 1906.
6. *Buffalo (NY) Courier*, July 8, 1906.
7. Glenn H. Curtiss and Augustus Post, *The Curtiss Aviation Book* (New York: Frederick A. Stokes, 1912), part 1, chap. II, 25.
8. *Hammondsport (NY) Herald*, September 12, 1906.
9. "A Motor Wind Wagon," *Popular Mechanics*, November 1906, 1106.
10. *Hammondsport (NY) Herald*, July 7, 1907.
11. *Hammondsport (NY) Herald*, December 12, 1906.
12. *Hammondsport (NY) Herald*, December 26, 1906.
13. *Hammondsport (NY) Herald*, September 12, 1906.
14. *Hammondsport (NY) Herald*, December 25, 1907.
15. "The Second Exhibition of the Aero Club of America," *Scientific American*, December 15, 1906, 448–49.
16. "The Famous Curtiss Mile," *Hammondsport (NY) Herald*, February 6, 1907; "Bright Future for Airship," *Inter Ocean* (Chicago), February 3, 1907, 53; and *Hammondsport (NY) Herald*, July 3, 1907.

Chapter 9: The Great V-8 and the Ormond Beach Speed Trials

1. "Kleckler, Henry, working on the V-8 motorcycle" (unpublished document), 1962, Kleckler, Henry file #289.0, G. H. Curtiss Museum, Hammondsport, NY.
2. Leonard C. Waters, "Tank Waters Recollections on the V-8 Motorcycle" (unpublished document), 1962, Waters, C. Leonard file #531.0, G. H. Curtiss Museum, Hammondsport, NY.
3. Letter to Floyd Clymer from James T. Sullivan, automotive editor, *Boston Globe*, in *Floyd Clymer's Motor Scrapbook*, vol. 2 (1944).
4. Waters, "Tank Waters Recollections on the V-8 Motorcycle."
5. "The Fastest and Most Powerful American Motor Bicycle," *Scientific American*, February 9, 1907, 128.
6. "Doings on the Beach," *Bicycling World & Motorcycle Review*, February 2, 1907, 537.
7. "Were 'Kings' of the Carnival," *Bicycling World & Motorcycle Review*, January 26, 1907, 509.
8. "The Accident to the Stanley Steam Racer," *Scientific American*, February 9, 1907, 128.
9. "Doings on the Beach," *Bicycling World & Motorcycle Review*, February 2, 1907, 537.
10. "Entertains Employees," *Hammondsport (NY) Herald*, February 6, 1907.
11. *Hammondsport (NY) Herald*, November 20, 1907; and *Hammondsport (NY) Herald*, December 11, 1907.

12. *Hammondsport (NY) Herald*, October 30, 1907; and Photograph, Curtiss Motorcycle booth, New York Auto Show, October 1907, Curtiss Motorcycle Catalogs file #614.0, G. H. Curtiss Museum, Hammondsport, NY.

13. "The Motorcycle Exhibit at the Chicago Automobile Show," *Motorcycle Illustrated*, January 1908, 13.

14. Handbill, Dayton Dry Goods Store, Minneapolis, MN, Curtiss V-8 motorcycle file # 624.0, G. H. Curtiss Museum, Hammondsport, NY.

15. Photograph, Chas. H. Wakeman & Son on V-8, Curtiss V-8 motorcycle file # 624.0, G. H. Curtiss Museum, Hammondsport, NY; Photograph, Arch Winney on V-8, Curtiss V-8 motorcycle file # 624.0, G. H. Curtiss Museum, Hammondsport, NY; and "Famous Curtiss Machine Shown," *Press and Sun-Bulletin* (Binghamton, NY), February 21, 1912, 9.

16. "Historic Motorbike Turns Up in Flexlume," *Curtiss Wright-er* (Buffalo, NY), February 10, 1943, 1.

17. Article from unknown Curtiss-Wright plant employee publication, V-8 Motorcycle file # 624.0, G. H. Curtiss Museum, Hammondsport, NY.

18. Smithsonian correspondence, Curtiss V-8 motorcycle file # 624.0, G. H. Curtiss Museum, Hammondsport, NY.

Chapter 10: 1907

1. *Hammondsport (NY) Herald*, May 15, 1907.
2. *Hammondsport (NY) Herald*, June 5, 1907.
3. *Hammondsport (NY) Herald*, March 6, 1907.
4. *Bicycling World & Motorcycle Review*, December 1907.
5. Vol. VII, State of N.Y. Supreme Court Appellate Div., Fourth Dept., *Herring-Curtiss Co. vs. Glenn H. Curtiss* (1922), C. L. Waters Testimony, 4671.
6. *Hammondsport (NY) Herald*, July 3, 1907.
7. "An Airship Chauffeur," *American Aeronaut*, January 1908.

8. Hammondsport (NY) Herald, July 3, 1907.
9. *Hammondsport (NY) Herald*, December 11, 1907; and Stein, *The Motorcycle Industry in New York State*, 77.
10. *Hammondsport (NY) Herald*, June 26, 1907; and *Hammondsport (NY) Herald*, July 10, 1907.
11. *Hammondsport (NY) Herald*, July 3, 1907.
12. *Hammondsport (NY) Herald*, July 10, 1907.
13. *Cycle and Automobile Trade Journal*, September 1, 1907, 198.
14. *Hammondsport (NY) Herald*, August 7, 1907.
15. *Hammondsport (NY) Herald*, July 31, 1907.
16. *Motorcycle Illustrated*, March 1908, 68.
17. *Curtiss Motorcycles for 1909* (catalog) (Hammondsport, NY: G. H. Curtiss Mfg., 1909).
18. "Good Enough for Uncle Sam," *Hammondsport (NY) Herald*, April 10, 1907; and *Democrat & Chronicle* (Rochester, NY), September 8, 1907, 5.
19. *Hammondsport (NY) Herald*, February 13, 1907.
20. *Hammondsport (NY) Herald*, February 6, 1907.
21. "Substantial Enlargement of Curtiss Motor Cycle Plant," *Cycle and Automobile Trade Journal*, March 1, 1907, 208.
22. *Hammondsport (NY) Herald*, February 27, 1907.
23. "Organization of Air Ship Co. Completed," *Dayton (OH) Herald*, November 21, 1907, 2.
24. Photograph, Curtiss Motorcycle Catalogs file #614.0, G. H. Curtiss Museum, Hammondsport, NY.
25. *Hammondsport (NY) Herald*, November 20, 1907.
26. "Toledo Man Beats His Tutor, Baldwin, in Dirigible Balloon Race," *Weekly Corinthian* (Corinth, MS), October 31, 1907, 2.
27. *Hammondsport (NY) Herald*, October 16, 1907; *Hammondsport (NY) Herald*, October 30, 1907; and *Hammondsport (NY) Herald*, November 6, 1907.

28. "Another Aerial Record Is Gone," *Rochester (NY) Democrat & Chronicle*, November 27 1907, 3; and "Airship Speed 30 Miles an Hour," *St. Louis (MO) Globe-Democrat*, November 27, 1907, 1.
29. *Hammondsport (NY) Herald*, November 27, 1907.
30. *Hammondsport (NY) Herald*, July 17, 1907.
31. Waters, "Tank Waters Recollections on the V-8 Motorcycle."

Chapter 11: The Aerial Experiment Association

1. G. H. Curtiss, "Accomplishments of the Aerial Experiment Association," report by G. H. Curtiss, director of experiments. 1909.
2. *Hammondsport (NY) Herald*.
3. Charles A. Champlin, "An Early Evening Jaunt, June 1908" (unpublished document), Champlin, Charles D. & A. family file #107.25, G. H. Curtiss Museum, Hammondsport, NY.

Chapter 12: 1908

1. *New York Times*, June 17, 1908.
2. *Hammondsport (NY) Herald*, September 16, 1907.
3. *Curtiss Motors* (catalog) (Hammondsport, NY: G. H. Curtiss Mfg., 1908).
4. Vol. VIII, State of N.Y. Supreme Court Appellate Div., Fourth Dept., *Herring-Curtiss Co. vs. Glenn H. Curtiss* (1922), G. H. Curtiss Testimony, 4987.
5. *Motorcycle Illustrated*, July 1, 1908, 3–4.
6. "The Scientific American Trophy for Flying Machines Heavier Than Air," *Scientific American*, September 14, 1907, 191.
7. "The Baldwin Airship," *Aeronautics*, October 1908.
8. "German Government Buys Airship in Hammondsport," *Hammondsport (NY) Herald*, December 2, 1908,
9. *Motorcycle Illustrated*, September 15, 1908, 1.
10. *Motorcycle Illustrated*, July 1, 1908, 8.
11. *Curtiss Motorcycles* (catalog) (Hammondsport, NY: Curtiss Motorcycle, 1911).
12. *Buffalo (NY) Commercial*, July 15, 1910.
13. *Bicycling World and Motorcycle Review*, October 31, 1908.
14. *Bicycling World and Motorcycle Review*, November 7, 1908.

Chapter 13: 1909

1. "Mr. Cornish Makes Some Notes," *Motorcycle Illustrated*, December 15, 1909, 6.
2. *Curtiss Motorcycles for 1909* (catalog) (Hammondsport, NY: G. H. Curtiss Mfg., 1909); *Hammondsport (NY) Herald*, January 6, 1909.
3. Vol. VII, State of N.Y. Supreme Court Appellate Div., Fourth Dept., *Herring-Curtiss Co. vs. Glenn H. Curtiss* (1922), H. C. Genung Testimony, 4786.
4. *Curtiss Motorcycles for 1909*.
5. *Curtiss Motorcycles* (catalog) (Hammondsport, NY: Curtiss Motorcycle, 1911).
6. "Injured on Motor Ice Boat," *Philadelphia Inquirer*, February 21, 1909, 9.
7. "Factory in Hammondsport," *Hammondsport (NY) Herald*, March 10, 1909.
8. *Motorcycle Illustrated*, July 1, 1909, 26.
9. *Motorcycle Illustrated*, December 1, 1909, 37.
10. *Hammondsport (NY) Herald*, April 14,1909.
11. *Motorcycle Illustrated*, August 1, 1909.
12. *Catalog of the Erie Motorcycles Motors and Accessories 1909* (Hammondsport, NY: Motorcycle Equipment & Supply, 1909).
13. *Hammondsport (NY) Herald*, November 20, 1907.

Chapter 14: The Marvel

1. Marvel Motorcycle Co. to Lena P. Curtiss, corporation deed, November 22, 1912, Steuben Co., NY, Marvel file, G. H. Curtiss Museum, Hammondsport, NY.
2. *Motorcycling Magazine*, February 1910.
3. Vol. VII, State of N.Y. Supreme Court Appellate Div., Fourth Dept., *Herring-Curtiss Co. vs. Glenn H. Curtiss* (1922), H. Kleckler Testimony, 4705
4. Vol. VII, State of N.Y. Supreme Court Appellate Div., Fourth Dept., *Herring-Curtiss Co. vs. Glenn H. Curtiss* (1922), H. G. Genung Testimony, 4728
5. *Curtiss Motorcycles* (catalog) (Hammondsport, NY: Herring-Curtiss, 1910).
6. *The 1911 Marvel Motorcycle* (advance flyer) (Hammondsport, NY: Marvel Motorcycle, 1911).
7. Vol. VII, State of N.Y. Supreme Court Appellate Div., Fourth Dept., *Herring-Curtiss Co. vs. Glenn H. Curtiss* (1922), H. Kleckler Testimony, 4423.
8. Ibid., 4720 and 5534.
9. *Motorcycle Illustrated*, December 19, 1909.
10. Vol. VII, State of N.Y. Supreme Court Appellate Div., Fourth Dept., *Herring-Curtiss Co. vs. Glenn H. Curtiss* (1922), C. L. Waters Testimony, 4676–78.

Chapter 15: 1910

1. Vol. VIII, State of N.Y. Supreme Court Appellate Div., Fourth Dept., *Herring-Curtiss Co. vs. Glenn H. Curtiss* (1922), V. E. Johns Testimony, 5699.
2. Vol. VII, State of N.Y. Supreme Court Appellate Div., Fourth Dept., *Herring-Curtiss Co. vs. Glenn H. Curtiss* (1922), H. Kleckler Testimony, 4712.
3. Ibid.
4. "Kleckler, Henry" (unpublished document), 1962, Kleckler, Henry file #289.0, G. H. Curtiss Museum, Hammondsport, NY.

5. Vol. VII, State of N.Y. Supreme Court Appellate Div., Fourth Dept., *Herring-Curtiss Co. vs. Glenn H. Curtiss* (1922), H. C. Genung Testimony, 4902–03.
6. Vol. VII, State of N.Y. Supreme Court Appellate Div., Fourth Dept., *Herring-Curtiss Co. vs. Glenn H. Curtiss* (1922), H. Kleckler Testimony, 4483–84.

Chapter 16: 1911

1. "The Marvel for 1911," *Cycle and Automobile Trade Journal*, January 1, 1911, 306.
2. *The 1911 Marvel Motorcycle* (advance flyer) (Hammondsport, NY: Marvel Motorcycle, 1911).

Chapter 17: 1912: The End of an Era

1. *Curtiss Fuselage* (Buffalo, NY), January 18, 1919.
2. *Curtiss Motorcycles* (catalog) (Hammondsport, NY: Curtiss Motorcycle, 1911).
3. Quote by G. H. Curtiss Museum curator Richard Leisenring.

INDEX

GLENN H. CURTISS MUSEUM OF LOCAL HISTORY

The Glenn H. Curtiss Museum of Local History, is located in the Village of Hammondsport, Town of Urbana, New York, at the southern tip of Keuka Lake. It was founded in 1961 by Otto Kohl, who was an associate and former employee of Curtiss. Kohl strongly believed that Glenn Curtiss was quietly being forgotten and wanted to preserve his legacy for future generations. He enlisted the help of members of the Curtiss family, such as Glenn Curtiss, Jr. and Carl Adams, along with Blanche Stuart Scott, America's first female aviator—taught by Glenn Curtiss personally—and the local community. The dream became a reality when the newly formed museum was granted a charter by the New York State Board of Regents in 1962. Today, 60+ years later, the institution is world renowned and hosts well over 25,000 visitors annually.

The official mission statement reads:

The Glenn H. Curtiss Museum of Local History was formed to establish, conduct, operate and maintain a museum in the Village of Hammondsport, Steuben County, New York, for the display of items relating to and significant in the role of Hammondsport in early aviation and transportation, particularly the contributions made to aviation by Glenn H. Curtiss, a resident of Hammondsport. The museum will display other items of historical significance which contribute to knowledge of Glenn H. Curtiss, Hammondsport and its regional history; conduct classes, exhibitions and observances which promote and stimulate the purposes for which the museum was formed.

The museum was originally located in the historic cobblestone elementary schoolhouse built in 1856 in the village proper where it operated for some thirty years. Eventually, the building—which also housed the village and town offices as well as the local library and police station—was becoming too cramped for growth. A more suitable building was located in 1992 to meet those needs by the museum's Board of Trustees. Located just outside the village on New York State Route 54, the 57,000-square foot building, known as the Henri Marchant Champagne Storage Warehouse, was very quickly obtained and converted to the current museum, with an additional upgrade in 2016, now in operation. The new building was originally part of the Urbana Wine Company, founded in 1865—a major connection to just one facet of the area's local history.

When the museum was first founded, it housed a very modest collection of Curtiss and local history artifacts which included five Curtiss aircraft, a large collection of early aviation engines, and two Curtiss motorcycles. Today, the museum has grown to exhibit the largest collection of Curtiss-related aircraft in the United States, which currently numbers twenty-five, both original and reproduction. It also is home to the largest collection of Curtiss and Hammondsport-built motorcycles displayed under one roof, totaling sixteen. Other exhibits include a large collection of vintage automobiles as well as various areas of the Curtiss legacy, past local industries, and the history of Hammondsport and Urbana. Also included is a large children's Science and Discovery area which offers many hands-on and STEM projects.

The Glenn H. Curtiss Museum is home to a large research archive pertaining to Glenn H. Curtiss, his family, businesses, early aviation, and local history. The archives are not open to the general public, but are available by appointment only through the museum's curator.

The museum also features a 75-seat theater, a large open area for special events which can be rented by the public, and a museum store. In addition to seeing the museum displays and exhibits, visitors are welcome to visit the Restoration Shop, talk with volunteer craftsman, and watch them work on historic aircraft.

For more information on the museum, exhibits, archives, admission rates, special exhibits, and events

that take place at the facility on an annual basis contact: info@curtissmuseum.org., see the web site at https:// glennhcurtissmuseum.org, or call 607-569-2160. Museum hours are May 1–October 31, 9 AM to 5 PM daily; November 1–April 30, 10 AM to 4 PM daily. The museum is closed on the following holidays: Easter, Thanksgiving, Christmas Eve, Christmas Day, and New Year's Day.

ABOUT THE AUTHOR

Richard Leisenring Jr. served as curator at the Glenn H. Curtiss Museum in Hammondsport, New York for twenty years until his retirement in 2023. He now serves as curator emeritus for the museum. He has been a professional historian and museum specialist for 45 years. Leisenring's other interests include research and collecting Civil War memorabilia, early photography, movie memorabilia, and Hopi Kachina figures. He lives in Bath, New York in the heart of the Finger Lakes with his wife and three cats.

Curtiss